Chahra BIDJOU-HAIOUR

Modified lipases. Approaches and prospects in sugar chemistry

Chahra BIDJOU-HAIOUR

Modified lipases. Approaches and prospects in sugar chemistry

lipasas recubiertas de tensioactivos

ScienciaScripts

Imprint

Cover image: www.ingimage.com

This book is a translation from the original published under ISBN 978-3-8416-2355-3.

Publisher:
Sciencia Scripts
is a trademark of
Dodo Books Indian Ocean Ltd. and OmniScriptum S.R.L publishing group

120 High Road, East Finchley, London, N2 9ED, United Kingdom
Str. Armeneasca 28/1, office 1, Chisinau MD-2012, Republic of Moldova, Europe
Managing Directors: Ieva Konstantinova, Victoria Ursu
info@omniscriptum.com

Printed at: see last page
ISBN: 978-620-8-41246-3

Contents

Introduction

The application of enzymatic catalysis to chemical processes reduces the use of harmful products in the environment, thereby cutting treatment costs. Enzymes offer a distinct advantage due to their specificity, biodegradability and limited formation of by-products. In recent years, there has been growing interest in the use of lipases for catalysis in organic media. By its very nature, lipase can have a significant influence on the evolution of chemical reactions when used as a biocatalyst. Its stability and activity in relation to the reagents can be improved by several processes.

One possible approach to stabilising enzymes is to immobilise them on a suitable matrix. Modification of enzymes changes their stability, enzymatic activity, pH optimum and reactivity with substrates. These changes depend on the source of the enzyme, the type of support and the modification method used.

The stability of enzymes can be improved by immobilising them on a suitable insoluble support, in particular via the amine or acid functions of the enzyme, which lead to the creation of physical or chemical bonds between the support and the enzyme [123] .

Immobilisation is the most widely used method for ensuring the stability of enzymes within an industrial process.

The synthesis of functional derivatives of molecules with surfactant potential, such as sugar esters, is an area of study that is constantly expanding. The development of green products and processes that are competitive with chemical products and processes is currently based on the concepts of green chemistry and sustainable development[2,3] . The cost of raw materials requires the use of simple, low-cost processes. This is the case with sugar-based biotensioactives obtained enzymatically[4] . The synthesis of these compounds using only renewable raw materials involves connecting a hydrophilic and a lipophilic part in the presence of an enzyme.

Thanks to their availability, sugars (as polar heads) and fatty acids (as apolar heads) have, over the last few decades, found their place in the esterification reaction [567] .

The best-known non-ionic surfactants of natural origin are sucroesters[6,7].

---

[1] E.N. Vulfson, P.J. Halling, H.L. Holland, Enzymes in nonaqueous solvents: methods and protocols, *Humana Press*, **2001**, 701.

[2] H. Oka, K. Aimono, S. Tsuyutani, T. Fujita, K. Hashiba. *Eur. Patent EP 492397, Kao Corp*. **1992**.

[3] H. Lüders, *Eur. Patent EP 252241, Hüls AG*. **1987**.

[4] H.A. El-Baz, A.M. Elazzazy, T.S. Saleh, M. Dourou, J.A. Mahyoub, M.N. Baeshen, H.R. Madian, and G. Aggelis. Enzymatic Synthesis of Glucose Fatty Acid Esters Using SCOs as Acyl Group-Donors and Their Biological Activities. *Applied Sciences*. **2021**, 11, 2700.

[5] N. Nicolas, B. Thierry, P. Daniel. Surfactants based on renewable substances. *L'actualité chimique*. **2000**. 70, 75.

[6] T. Raku, M. Kitagawa, H. Shimakawa, Y. Tokiwa. Enzymatic synthesis of hydrophilic undecylenic acid sugat esters and thier biodegradability. *Biotechnology letters*. **2003,** 161, 166.

[7] T. Raku, M. Kitagawa, H. Shimakawa, Y. Tokiwa. Enzymatic synthesis of trehalose esters having liphophilicity. *Journal of biotechnology*. **2003,** 100, 208.

These are biodegradable molecules that are harmless to health and the environment. They are totally biocompatible and can be used in all applications in direct contact with humans, such as food processing[8] , pharmaceuticals[9] , detergents, cosmetics[10] , food additives, etc. [11121314151617] . Depending on the carbohydrate and acyl fractions chosen, esters of sugars and fatty acids have been shown to have anti 1213,14,151617 oxidant, antimicrobial ", insecticidal and anti-tumour properties. They can be obtained chemically using alkaline catalysts[18] , but this strategy requires high temperatures and the use of dangerous solvents such as pyridine, which is not compatible with food applications. In addition, as all carbohydrate hydroxyl groups exhibit similar reactivity, the result generally mixtures of esters, with no control over the composition of the reaction product .[19]

Enzymes such as lipases, proteases and esterases are capable of catalysing the synthesis of sugar and fatty acid esters with a high degree of selectivity, yielding mono-esters directly without the need for additional protection and deprotection steps[20] . Among these, lipases are the enzymes most commonly used to catalyse the synthesis of such compounds. These enzymes are active in many organic solvents and at low temperatures . The enzymatic route has therefore been widely studied as a gentler, greener alternative for synthesising sugar esters from fatty acids[21] . It should be noted that global production of sugar esters is estimated at 6,000 t/year[22] . In the United States, the Food and Drug Administration (FDA) authorised the

---

[8] N. do Amaral Sampaio Neta, J.C. Sousa dos Santos, S. de Oliveira Sancho, S. Rodrigues, L. Rocha Barros Goncalves, L.R. Rodrigues, J.A. Texeira. Enzymatic synthesis of sugar esters and their potential as surface-active stabilizers of coconut milk emulsions. *Food Hydrocolloids*. **2012**, 27, 331.

[9] L. Schiefelbein, M. Keller, F. Weissmann, M. Luber, F. Bracher, W. Friess. Synthesis, characterization and assessment of suitability of trehalose fatty acid esters as alternatives for polysorbates in protein formulation. *European journal of pharmaceutics and biopharmaceutics*. **2012**, 76, 350.

[10] N.R. Kahn, V.K. Rathod. Enzyme catalyzed synthesis of costemic esters and its intensification: a review. A review. *Process Biochemistry*. 2015, 50, 1806.

[11] C. N. Mulligan. Environmental applications for biosurfactants. *Environ. Pollution*. **2004**, 133, 198.

[12] S.W. Hsieh, M.R. Lee, C.W. Tsai, L.Y. Lai, T. Yeh, C.W. Hsieh, T.J. Yang, S.W. Chang, Enzymatic synthesis, purification and identification of bioactive trehalose ester derivatives for health applications. *Food Bioprod. Process*. **2015**, 95, 172.

[13] M. Habulin, S. Sabeder, Z. Knez. Enzymatic synthesis of sugar fatty acid esters in organic solvent and in supercritical carbon dioxide and their antimicrobial activity. *The Journal of Supercritical Fluids*. **2008,** 45, 345.

[14] M. Ferrer, J. Soliveri, F.J. Plou, N. Lopez-Cortes, D. Reyes-Duartes, M. Christensen, J.L. Copa-Patino, A. Ballesteros. Synthesis of sugar esters in solvent mixtures by lipases from Thermomyces lanuginosus and Candida antarctica B, and their antimicrobial properties, *Enzyme and Microbial Technology*. **2005,** 36, 398.

[15] T. Watanabe, S. Katayama, M. Matsubara, Y. Honda, M. Kuwahara. Antibacterial carbohydrate monoesters suppressing cell growth of Streptococcus mutans in the presence of sucrose. *Current microbiology*. **2000,** 41,213.

[16] J.K. Peterson, H.F. Harrison, O.T. Chortyk. Effects of various synthetic sucrose esters on weed seed germination and crop growth: structure - activity and dose - response relationships. *Journal of Agricultural and Food Chemistry*. **1997,** 45, 4837.

[17] S. Okabe, M. Suganuma, Y. Tada. Dissacharide esters screened for inhibition of tumor necrosis factor - alpha release are new anti-cancer agents. *Japanese journal of cancer research*. **1999,** 90, 676.

[18] C.C. Akoh, B.G. Swanson. Synthesis and properties of alkyl glycoside and stachyose fatty acid polyesters. *Journal of the American Oil Chemists ' Society*. **1989,** 66, 1301.

[19] G. Paggiola. Biocatalysis in bio-derived solvents: an improved approach for medium optimisation, *Green Chemistry*. **2014,** 16, 2110.

[20] P. Fickers, J. Destain, P. Thonart. Lipases are atypical hydrolases: main characteristics and applications. *Biotechnology agronomy, society and environment*. **2008,** 12, 130.

[21] A. Dounia, G. Etienne, N. Stéphane, 6-O-glucose palmitate synthesis with lipase: Investigation of some key parameters. *Molecular Catalysis*. **2018,** 460, 68.

[22] A. K. Chandel, R. Rudravaram, L. V. Rao, P. Ravindr, M. L. Narasu. Industrial enzymes in bioindustrial sector development: an Indian perspective. *Journal. Commercial Biotechnology*. **2007,** *13*, 283.

use of sugar esters in 1983 as emulsifiers and texture agents in foodstuffs or as a protective coating for preserving fruit[23] . In Europe, the dietary consumption of sucroesters (E473) was authorised in 1989, with a limit of 25mg/kg/day.

Demand for sugar esters is expected to increase as new, more selective processes are discovered for manufacturing compounds that meet economic and environmental requirements .[24]

The application of enzymatic catalysis to chemical processes reduces the use of products that are harmful to the environment, thereby cutting processing costs. In this case, lipases offer a distinct advantage due to their specificity, biodegradability and limited formation of by-products. One possible approach to stabilising lipases in organic solvents is to immobilise them on a suitable matrix adsorption onto a solid support [25][26][27] and coating with non-ionic surfactants[26,27]. This changes their stability, their enzymatic activity and their reactivity towards substrates. These changes depend on the source of the enzyme, the type of substrate and the method used. In the following, free *porcine pancreatic* lipase (PPL) and *Candida cylindracéa* lipase (CCL) are modified by the above processes. The supports used in the adsorption process are inorganic, namely celite and silica. These lipases are coated or encapsulated with sorbitan esters and non-ionic surfactants. The modified lipases were then tested in the enzymatic esterification reaction of D-glucose to compare their catalytic activities with those of the free enzyme and a commercial immobilised enzyme, *Candida antartica* lipase B (CAL B).

---

[23] S. Nakamura. Using sucrose esters as food emulsifiers. *Inform* **1997**, 8, 874.

[24] S. Cassel, C. Debaig, T. Benvegnu. P. Chaimbault, M. Lafosse, D. Plusquellec, P. Rollin. Original synthesis of linear, ranched and cyclic oligoglycerol standards. *European Journal of Organic Chemistry*. **2001**, 875, 1829.

[25] N. Bouzaouit, C. Bidjou-Haiour. Response surface methodological study of glucose laurate synthesis catalyzed by immobilized lipase from candida cylindracea. *Biological Forum-An Intern. J.* **2016**, 81, 420.

[26] S. Okazaki, M. Goto, H. Wariishi, H. Tanaka, S. Furusaki. Characterization and catalytic property of surfactant-laccase complex in organic media. *Biotechnology Progress*, **2000**, 16, 583.

[27] V. Dandavate, D. Madamwar. Novel approach for the synthesis of ethyl iso valerate using surfactant coated *Candida rugosa* lipase immobilized in microemulsion based organogels. *Enzyme and microbial technology*. **2007**, 41,265.

CHAPTER 1

Immobilisation of enzymes

The development of new analytical methods for research and industry increasingly requires the use specific enzymes. It is therefore essential to find a method of stabilising enzymes against denaturation.

In fact, work enzyme immobilisation only began in the 1960s, when enzymes became widely available.[28]

For an enzymatic reaction to take place under the best possible conditions, it must be stable. In functional terms, the stability of an enzyme often limits its practical application in medicinal and biotechnological processes. One approach increasing the stability of these enzymes immobilisation on a suitable matrix, to preserve the active three-dimensional structure.

The enzyme may have a more restrictive conformation in its immobilised state, causing a reduction in its activity or blocking its catalytic site. Confinement matrices must therefore allow free diffusion of the substrate towards the active site and favour a sufficiently free spatial conformation to preserve its biological activity. If these conditions are met, the advantages of immobilisation are limited release caused by rinsing, a more stable enzyme and longer life.

From an industrial point of view, it was observed that immobilised biocatalysts were more stable and in some cases could induce changes enzymatic activity, pH optimum and substrate affinity.

Secondly, immobilising the enzyme means that it can either be used in a continuous reactor, or recovered and reused (these two aspects can also be combined). It is therefore very simple to significantly increase the economy of an enzymatic process, especially as the cost of the enzyme is high.[29]

It is also important to remember that in living systems or in nature, enzymes practically only

---

28 [28] B. Eric, B. Jean-François. Enzymatic catalysis. *Techniques de l'ingénieur*, **1992,** 240, 1.

29 C. Didier, M. Pierre. Biocatalysis or enzymatic catalysis. *Techniques de l'ingénieur*, **2009**, 590, 1.

30 C. Blecker, M. Christophe, Immobilized lipases and their applications. *Biotechnology, Agronomy, Society and Environment*, **2008**, 12, 57

31 MF. Chaplin, C. Bucke. Enzyme Technology. *Cambridge University Press*, **1990.**

32 [29] I. Alexander, S. Manfred. Methods for the immobilization of lipases and their use for ester synthesis. *Journal of Molecular Catalysis*, **1997**, 3, 303.

33 D. Sumitra, R. Christena, R. Yamuna, R. Sriramulu. Enzyme immobilization: an overview on techniques and support materials. *Biotech*, **2013**, 3, 1.

34 M. Eisenmenger, J. Reyes-De-Corcuera. High hydrostatic pressure increased stability and activity of immobilized lipase in hexane. *Enzyme and Microbial Technology*, **2009**, 45, 118.

35 Z. Dijkstra, R. Merchant, J.T.F. Keurentjes. Stability and activity of enzyme aggregates of *CALB* in supercritical CO_2. *Journal of Supercritical Fluids*, **2007**, 41, 102.

act in an immobilised state, such as on the surface of the membrane structures of a cell or a cellular organelle. Immobilising an enzyme therefore makes it easier to control and determine the effect of the microenvironment on its catalytic activity.[30]

Enzyme immobilisation is essential for the implementation of catalytic reactions on an industrial scale, and the various processes need to be mastered by manufacturers.

- The different immobilisation techniques

Immobilisation consists of adding various compounds to the enzyme, such as proteins, sugars, solid supports, etc. This results different types of enzyme preparations in powder or liquid form to give a coated enzyme or in the form of microgranules.

The choice of the type of immobilisation depends very much on the nature of the desired application. For example, covalent coupling of the enzyme is often used for the development of biosensors, while encapsulation of biomolecules is suitable for release into living cells in order to release the active molecule at its target during therapeutic treatments. In addition, the physico-chemical properties of the enzyme must be taken into account. Enzymes are sensitive to changes in temperature and pH, and exhibit optimal activity within a specific range of these parameters. It is therefore necessary to associate them with the matrix under conditions that are favourable to their preservation, i.e. to develop hybrid materials under mild synthesis conditions (mild chemistry processes). The different modes of enzyme immobilisation and the different containment matrices used.[31]

There are four main immobilisation methods involving different physical and chemical processes, as shown in Figure 1.

- Immobilisation by adsorption.
- Immobilisation by covalent bond.
- Immobilisation by inclusion.
- Immobilisation by encapsulation.

---

[36] [30] D. Combes, P. Monsan. Biocatalysis or enzymatic catalysis. *Techniques de l'ingénieur*, **2009,** 11, 1.

[37] M.F. Chaplin, C. Bucke. Enzyme Technology. *Cambridge University Press*, **1990.**

[38] [31] J. M. Guisan. Immobilization of Enzymes and Cells. *Methods Biotechnol*, **2006**, 22.

U. Hanefeld, L. Gardossi and E. Magner, Understanding enzyme immobilisation. *Chemical Society Reviews* **2009**, 38, 468.

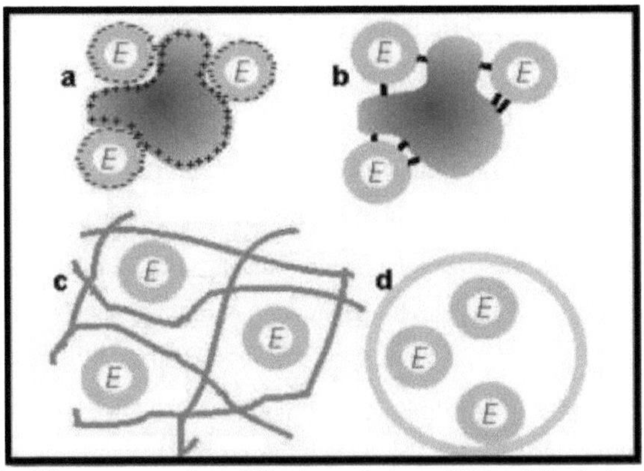

Figure 1- *The different enzyme immobilisation techniques* (a) Immobilisation by adsorption; (b) Immobilisation by covalent bonding; (c) Immobilisation by inclusion; (d) Immobilisation by encapsulation .[32]

a. Immobilisation by adsorption

Adsorption enzymes onto insoluble matrices is the simplest method of immobilisation. The procedure incubating the protein with the adsorbent under appropriate conditions (pH, temperature). The hybrid material is recovered after filtration or centrifugation. The interactions that exist between the enzyme and the adsorbent are weak interactions of the electrostatic or Van der Waals type. However, the adsorption process is reversible and can, under certain conditions of use, lead to desorption of the biomolecule and therefore to alteration of the material. However, of the different immobilisation methods, adsorption is the one that induces the least modification of the material. Another important advantage of enzyme immobilisation is that the immobilised enzymes can exhibit properties that can be exploited in reactions carried out in non-aqueous media. Typically, native enzymes are active in aqueous media. However, when immobilised, enzymes can have their catalytic properties modified to enable them to retain their catalytic activity under non-aqueous conditions. This is important because such enzymes can be used for substrate transformations that can only be carried out in organic solvents [333435] . On the other hand, immobilised enzymes may exhibit catalytic properties in organic solvents that differ from those observed in aqueous media .[36]

---

[32] M.F. Chaplin, C. Bucke. Enzyme Technology. *Cambridge University Press*, **1990.** conformation of the enzyme, making it a preferred technique for immobilising enzymes[41,42].

[33] M. M. Zheng, Y. Lu, L. Dong, P. M. Guo, Q. C. Deng, W. L. Li, Y. Q. Immobilization of Candida rugosa lipase on hydrophobic/strong cation-exchange functional silica particles for biocatalytic synthesis of phytosterol esters. *Bioresource technology*, **2012**, 115, 141.

[34] A. Bastida, P. Sabuquillo, P. Armisen, R. Fernandez-Lafuente, J. Huguet and J. M. Guisan, A single step purification, immobilization, and hyperactivation of lipases via interfacial adsorption on strongly hydrophobic supports. *Biotechnology and bioengineering*, **1998,** 58, 486.

[35] V, Iyer, S. Padma, L. Ananthanarayan. Enzyme stability and stabilization-aqueous and non-aqueous environment. *Process biochemistry*. **2008,** 43, 1019.

[36] Klibanov, M. Alexander. Improving enzymes by using them in organic solvents. *Nature*, **2001**, 409, 241.

A wide variety of adsorbent media are used, both in terms of their chemical structure and their physical properties (density, granulometry, porosity). They can be classified into two categories .[37]

> **Inorganic supports:** aluminosilicates, silica, porous glass, metal oxides, activated carbon, clays (bentonite, kaolin), calcium carbonate, etc.

> **Organic substrates:** these can be natural (collagen, cellulose, starch, agarose) or synthetic (acrylic polymers, methacrylic polymers, polystyrene, copolymers, ion exchange resins).

The quantities of enzymes immobilised by adsorption are extremely variable and obviously depend on the nature of the support used. Even if this process remains simple in terms of both technology and implementation, several parameters influence efficiency of adsorption.

- Enzyme concentration
- The specific area available per unit mass of support
- Contact time
- The reaction environment

b. Immobilisation by covalent bonding

Immobilisation by covalent bonding was developed with the aim of obtaining very strong bonds between enzymes and the support[38] . This is the best way avoiding any risk of releasing enzyme molecules into the environment. To achieve immobilisation by covalent bonding, it is necessary to first activate the attachment support or the enzyme itself. This involves involving the reactive functional groups ($-NH_2$, -OH, -SH, $-CO_2H$) of amino acids such as lysine, histidine, tyrosine, glutamate or aspartate, present on the surface of the enzyme molecule, which has very limited chemical reactivity. The covalent bond must be established under mild conditions to avoid any denaturation of the enzyme and prevent any functional group from being altered.

Two groups of immobilisation methods can be distinguished by covalent bond: on an insoluble support or by cross-linking .[39]

- **Immobilisation by cross-linking:** the principle is to bridge the enzyme molecules with each other or with other proteins using a cross-linking agent such as glutaraldehyde to obtain insoluble macrostructures.
- **Supported immobilisation:** bringing the enzyme into contact with an activated support carrying groups sufficiently reactive to react with those of the enzyme under conditions compatible

[37] E. Brown, B. Jean-François. Enzymatic catalysis. Techniques de l'ingénieur, **1992**, 240, 1.

[38] L. Ren, J. He, S. Zhang, D.G. Evans, X. Duan, Immobilization of penicillin G acylase in layered double hydroxides pillared by glutamate ions. *Journal of Molecular Catalysis B: Enzymatic*,, **2002**, 183, 11,

[39] D. Combes, P. Monsan. Biocatalysis or enzymatic catalysis. *Techniques de l'ingénieur.* **2009**,11, 590.

with the preservation of catalytic activity.

The immobilisation of enzymes by this method of preparation can be adapted to a wide variety of matrices and binding agents, in particular for the development of biosensors. Tyrosinase (PPO)[40] was immobilised on a polymer film by covalent bonds between the enzyme and a conductive polymer, poly-(3-aminopropyl pyrrole), to develop a biosensor for phenol, an environmentally toxic molecule. The bond is created by peptide coupling between the free amine groups present on the surface of the polymer and the carboxylic acid groups of the enzyme (Figure 1), for one hour in a phosphate buffer solution (0.1 M, pH=7.2). These results show that the polymer can be as an immobilisation platform to develop modified electrodes for biosensors. In addition, the low cost and simplicity of the immobilisation method offer further advantages for this type of use.

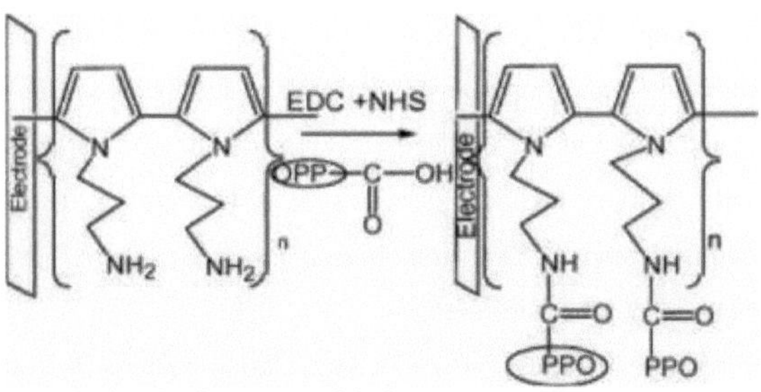

Diagram 1- *Stages involved in the formation of the biosensor*

c. Inclusionary immobilisation

The principle of inclusion is to hold the enzyme captive in the matrix of a polymer or in a microcapsule. Several polymers are frequently used, such alginate, chitosan, polyacrylamide gel or starch gel. To do this, the enzyme is dissolved in an aqueous solution of monomers that also contains a cross-linking agent. The co-monomers are then polymerised at low temperature to form a three-dimensional network in which the enzyme is retained. This immobilisation must, of course, leave the active site free and accessible to the substrate[49,50].

d. Immobilisation by encapsulation

This technique involves enclosing the enzymes in a permeable, flexible material that partially isolates and protects them from the outside environment. The catalyst can be confined in a polymer gel or inside a capsule [414243] .

---

[40] K. R. Kaneto, A new tyrosinase biosensor based on covalent immobilization of enzyme on N-(3-aminopropyl) pyrrole polymer film. *Current Applied Physics*, **2005**, 5, 178.

[41] E. Brown, B. Jean-François. Enzymatic catalysis. *Techniques de l'ingénieur*, **1992**, 240, 1.

[42] B. Christophe. Immobilised lipases and their applications, *Biotechnology, Agronomy, Society and Environment*, **2008**, 12, 68.

[43] D. Sumitra, R. Christena, R. S. R. Yamuna. Enzyme immobilization: an overview on techniques and support materials. *Biotech,* **2013**, 3, 9.

Immobilisation here is physical and not chemical, unlike immobilisation by covalent bonding, for example. The matrix must allow the diffusion of small molecules only (in particular the substrates and products of the reaction) but under no circumstances must it allow the diffusion of larger molecules, including enzymes. These matrices can be inorganic (silica gels), organic (nafion), polymeric (polyacrylamide, polyurethanes) or composite (carbon paste).

This immobilisation method has many advantages: it is economical, easy to implement and can applied to a large number of enzymes[52,53,54].

Chen *et al.* trapped the lipase [44454647] in an organic-inorganic sol-gel hybrid matrix prepared by polymerising tetramethoxysilane ($Si(OCH_3)_4$) with alkyltrimethoxysilane ($RSi(OCH_3)_3$). Optimum preparation conditions determined in order to obtain the best dispersion of the enzymes. Lipophilic interactions between the hydrophobic domains of the enzyme and the functionalized alkyl groups in the gel ($RSi(O^-)_3$) play a key role in fixing the biomolecule in a more active conformation. For comparison, free lipase was lyophilised from a buffer solution (0.1M) at pH= 7.2 and stored at -20°C. The percentage of immobilised protein was 95% and the resulting specific activity of the lipase was 59 times greater than that of the non-immobilised lyophilised lipase. The immobilised enzyme showed greater stability to temperature and storage than the free enzyme.

Livage *et al*[48] demonstrated the feasibility of encapsulating bacteria in a silica gel produced by a sol-gel process (Figure 2). This work showed that cellular integrity was maintained in a wet gel and that enzymatic activity was enhanced by dispersion of the bacteria in the matrix. However, drying the gel resulted in a loss of enzymatic activity.

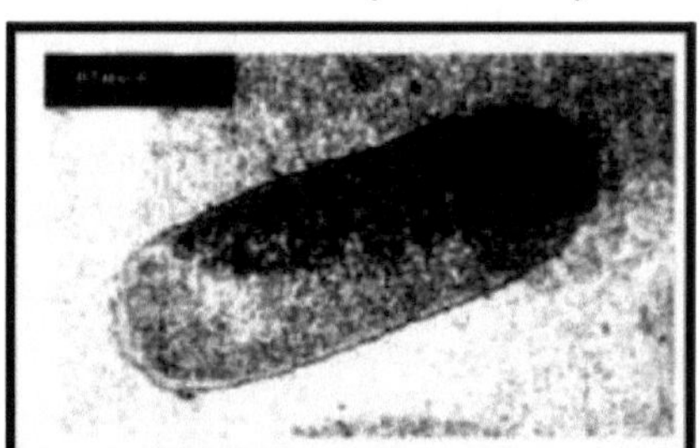

Figure 2- *Bacteria trapped in a moist gel*

[44] G. Iqbal, A. Ballesteros. Bioencapsulation within synthetic polymers (Parti): sol-gel encapsulated biologicals. *Tibtech,* **2000**, 18, 296.

[45] M. Maciej, A. Krywko-Cendrowska, P. Krysinski, J. Rogalski. Encapsulation of laccase in a conducting polymer matrix: A simple route. *Synthetic Metals*, **2009**, 159, 1738.

[46] L. Zhang, Y. Jiang, Sh. Jiafu, S. Xiaohui, Li Jian, Biomimetic polymer-inorganic hybrid microcapsules for yeast alcohol dehydrogenase encapsulation. *Reactive & Functional Polymers*, **2008**, 68, 151.

[47] J.P. Chen, W.S. Lin, Sol-gel powders and supported sol-gel polymers for immobilization of lipase in ester synthesis. *Enzyme and Microbial Technology*, **2003**, 32, 801.

[48] S. Fennouh, S. Guyon, C. Jourdat, J. Livage, C. Roux, Encapsulation of bacteria in silica gels. *Comptes Rendus de lAcadémie des Sciences-Series IIC-Chemistry*, **1999**, 625.

Immobilisation can also be carried out in spherical capsules made of materials whose size can be adjusted from nanometres to micrometres depending on the size of the biomolecules to be immobilised. The biomolecule can be confined, for example, in alginate beads, liposomes or gold particles. The preparation of these materials is highly targeted, and mainly aimed at medical research for the controlled release of therapeutically active molecules into target cells.

Transmission electron microscopy can be used to check the effectiveness of enzyme or cell trapping inside the capsules. For example, in the case immobilisation of C2C12 Myoblast cells in an alginate microcapsule (Figure 3).

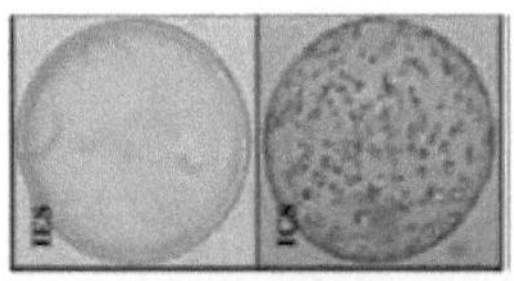

a)* *b)

***Figure 3** - Capsule a) without and b) with cells*[57]

The final example concerns the immobilisation of horseradish peroxidase (HRP) in hollow gold particles. The preparation method involves i) the formation of reverse micelles encapsulating the HRP, ii) the growth of gold particles on the surface of the micelles and then iii) the solubilisation of the organic micelles leaving the hollow gold nanoparticles containing the enzyme. Figure 4 shows the reaction diagram for this type of synthesis. Diffusion of the small substrate molecules through the pores of the gold particles is effective. The enzyme remains chemically active, allowing catalytic reactions to take place. [4950]

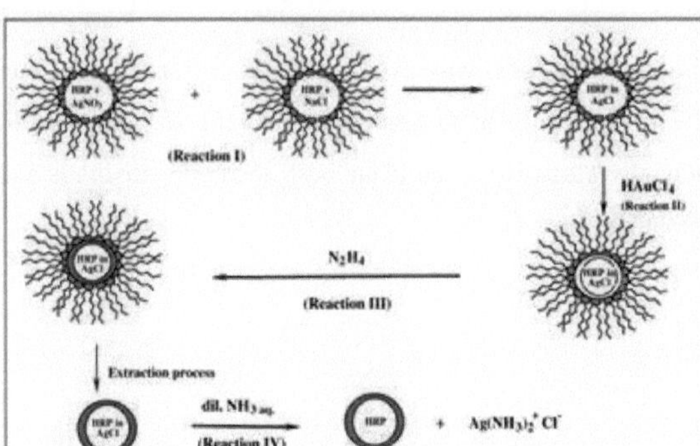

***Figure 4**- Enzyme encapsulation in a hollow gold particle*[51]

[57] G. Orive, R. M Hernandez, A.R. Gascon, M. Igartua, J. P. Pedraz, Microcapsules prepared with different biomaterials to immobilize GDNF secreting 3T3 fibroblasts. *International journal of pharmaceutics*, **2003**, 259, 68.

[58] R. Kumar, A. N Maitra, P. K. Patanjali *et al.* Hollow gold nanoparticles encapsulating horseradish peroxidase. *Biomaterials*. **2005**, 26, 6743.

[51] R. Kumar, A. N Maitra, P. K. Patanjali *et al.* Hollow gold nanoparticles encapsulating horseradish peroxidase. *Biomaterials*. **2005**, 26,

Enzymes can also be made more stable by using surfactants (micelle formation)[52] as a containment agent. This is the coating technique.
Enzyme coating is a technique used to protect proteins from the stresses of the reaction environment (pressure, solvent, mechanical shear, etc.). It can generally be achieved by trapping or encapsulating the enzymes in a surfactant or polymer matrix. A wide variety of polymers are used, from sodium alginates to silica polymers.
Most of the preparation procedures are relatively simple and are carried out under mild conditions, often at room temperature, and mainly involve mixing an aqueous solution of the lipase with an alcoholic solution of the surfactant using ultrasound. The result is an enzyme confined in a micelle (Figure 5). The most critical issue in the preparation surfactant-coated lipases is the choice of surfactant. It should be noted that some of them contain ester functions, which means that there is a risk that they will be degraded by the lipases used. From this point of view, surfactants without ester functions are a safer choice, especially when the reaction lasts over time. Sorbitan esters have frequently been used to prepare surfactant-coated lipases , with sorbitan monostearate (span 60) being the most popular.

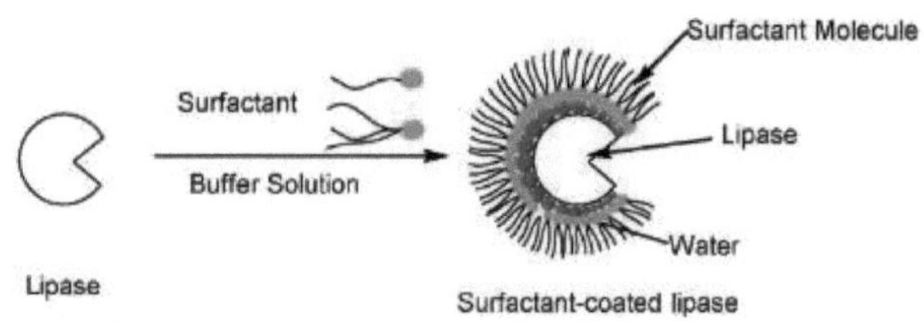

Figure 5- *Enzyme coating*

To date, immobilisation is the best way avoiding the negative impact of reaction media on enzyme structure. The simplicity of the processes and the associated mild conditions also make them commercially attractive, even if the cost of the precursors is added to that of the enzymes to be immobilised.

---

[52] P.V. Iyer, L. Ananthanarayan, Enzyme stability and stabilization: Aqueous and nonaqueous environment, *Process Biochemistry*, **2008,** 43, 1019.

CHAPTER 2

Modification of lipases

Recently, research into enzymatic reactions in organic media has attracted more attention as a new organic synthesis procedure [53545556575859] . Lipases are also widely used in various organic syntheses, particularly esterification[61,62]. They are reported have high stability and activity and are conventionally used in aqueous and organic media[63,64]. However, the usefulness of these enzymes in industrial processes is limited by their tendency to be denatured and inactivated when exposed to organic solvents. Approaches to overcome this problem have been successfully developed[65,66]. Immobilisation of free lipase on suitable supports by various processes such as adsorption [6061626364]or encapsulation in non-ionic surfactants[68,69] are techniques that generally give good results and a considerable increase in catalytic activity[70,71.] They also make it possible to prolong the existence of these enzymes. The most significant advantages of modified enzymes are their simple preparation procedures and their good solubility in a wide range of organic solvents.

a. Modification of lipases by immobilisation (adsorption)

Porcine pancreatic lipase and *Candida cylindracéa* lipase are immobilised on celite and silica, which are microporous supports, by the adsorption process[65] . Physical adsorption is the simplest method of immobilising enzymes. It involves an enzyme solution, generally aqueous, simply being brought into contact with a selected material used as a support. Adsorption will occur provided the enzyme-support interactions are sufficiently strong. In the case of lipases, hydrophobic interaction is the most common, but ionic interactions with ion

---

[53] N. Kamiya, M. Goto, F.Nakashio. Surfactant-Coated Lipase Suitable for the Enzymatic Resolution of Menthol as a Biocatalyst in Organic Media. *Biotechnologie. Progress*. **1995**, 11, 270.
[54] K.P. Dhake, D.D. Thakare. B.M. Bhanage. Lipase: A potential biocatalyst for the synthesis of valuable flavour and fragrance ester compounds. *Flavour Fragr. J.* **2013**, 28, 71.
[55] M.J. Ferrer, J.P. Francisco, N. López-Cortés, D. Reyes-Duarte, M. Christensen, J.L. Copa-Patiño, A. Ballesteros. Synthesis of sugar esters in solvent mixtures by lipases from *Thermomyces lanuginosus* and *Candida antarctica* B and their antimicrobial properties. *Enzyme and microbial technology*. **2005**, 36, 391.
[56] A.M. Klibanov. Improving enzymes by using them in organic solvents. *Nature*, **2001**, 108, 241.
[57] A. Zaks, A.M. Klibanov, Enzyme-catalyzed processes in organic solvents, *Proceedings of the National Academy of Sciences*, **1985**, 82, 3192.
[58] D. Feng-Xia, L. Zhang, T. Xiao-Zuo, Ch. Hong-Bing, W. Xiu-Li and W. Yu-Zhong, Ionic liquid coated lipase: Green synthesis of high molecular weight poly (1,4-dioxan-2-one). *Journal of Molecular Catalysis B: Enzymatic*. **2012**, 77, 46.
[59] T. Mori, S. Kishimoto, K. Ijiro, A. Kobayashi and Y. Okahata. A Lipid-Coated Lipase as an Efficient Hydrolytic Catalyst in the Two-Phase Aqueous-Organic System, *Biotechnology and bioengineering*. **2001**, 76, 157.
[60] N. Bouzaouit, C. Bidjou-Haiour. Response Surface Methodological Study of Glucose Laurate Synthesis Catalyzed by Immobilized Lipase from *Candida cylindraceaBiological Forum-An International Journal*. **2016**, 81, 420.
[61] V. Dandavate and D. Madamwar, Novel approach for the synthesis of ethyl isovalerate using surfactant coated *Candida rugosa* lipase immobilized in microemulsion based organogels, *Enzyme and microbial technology*. **2007**, 41,265.
[62] Y. Okahata and T. Mori, Lipid-coated enzymes as efficient catalysts in organic media, *Trinds in Biotechnolgy*. **1997**, 4, 15.
[63] N. Klai, C. Bidjou-Haiour. S. Bouquillon. D-Xylose-based surfactants: Synthesis, characterization and molecular modeling studies, *Comptes Rendus Chimie*. **2015**, 18, 599.
[64] A.M. Klibanov, Improving enzymes by using them in organic solvents, *Nature*, **2001**, 108, 241-246.
[65] N. Bouzaouit, C. Bidjou-Haiour. Response surface methodological study of glucose laurate synthesis catalyzed by immobilized lipase from candida cylindracea, *Forum-An International Journal*. **2016**, 81,420.

exchangers in the materials can also be useful [666768].

Since lipases spontaneously adsorb to hydrophobic surfaces in aqueous solutions to a greater degree than most other proteins, this method can be used for lipase immobilisation[74,75]. Interestingly, adsorption onto the support often leads to an increase in the catalytic activity of the lipase[69]. This is interpreted as being due to hydrophobic interactions between the support and part of the lipase molecule causing the lid to open to give the active conformation, in analogy with the interfacial activation caused by a hydrophobic substrate phase.

b. Modification of lipases by encapsulation (coating)

Another way of obtaining highly active enzyme preparations in organic media is to prepare protein (enzyme) coated microcrystals (PCMC)[70]. The same lipases (CCL and PPL) are then coated with non-ionic surfactants.

In this method, an aqueous enzyme solution is mixed with a concentrated solution of an excipient, which is normally an inorganic or organic compound, salt, sorbitant esters (span, twin...) or an amino acid. The mixed solution is slowly added to a water-miscible organic solvent to induce precipitation of the enzyme. The surface phenomenon means that the enzyme retains its activity and allows efficient mass transfer of substrates and products. Propanol or ethanol with a low water content are good solvents for preparing PCMCs, but other solvents can also be used[71]. Lipase PCMCs express activities up to 200 times higher than the same enzymes in free form[72]. In most of the studies reported, the enzyme molecules in PCMCs were not cross-linked. However, cross-linking with glutaraldehyde has been evaluated, producing protein-coated cross-linked microcrystals. This procedure might be expected to provide a biocatalyst that is more resistant to enzyme leakage, but surprisingly the catalytic activity was found to be higher than that of the corresponding PCMC without crosslinking for CRL and BCL in four different organic solvents [737475].

---

[66] M. M. Zheng, Y. Lu, L. Dong, P. M. Guo, Q. C. Deng, W. L. Li, Y. Q. Feng and F. H. Huang, Immobilization of Candida rugosa lipase on hydrophobic/strong cation-exchange functional silica particles for biocatalytic synthesis of phytosterol esters. *Bioresource technology*, **2012**, 115, 141.

[67] T. Gitlesen, M. Bauer and P. Adlercreutz, Biochim. Adsorption of lipase on polypropylene powder. *Biochimica et Biophysica Acta (BBA)-Lipids and Lipid Metabolism,* **1997**, 1345, 188.

[68] R. Fernandez-Lafuente, P. Armisen, P. Sabuquillo, G. Fernandez-Lorente and J. M. Guisan, Immobilization of lipases by selective adsorption on hydrophobic supports. *Chemistry and physics of lipids* ,**1998**, 93, 185.

[69] A. Bastida, P. Sabuquillo, P. Armisen, R. Fernandez- Lafuente, J. Huguet and J. M. Guisan, A single step purification, immobilization, and hyperactivation of lipases via interfacial adsorption on strongly hydrophobic supports. *Biotechnology and* **bioengineering1998**, 58, 486.

[70] M. Kreiner, B. D. Moore and M. C. Parker, Chem. Commun. Enzyme-coated micro-crystals: a 1-step method for high activity biocatalyst preparation, *Chemical communications.* **2001**, 1096-1097.

[71] M. Kreiner, J. F. A. Fernandes, N. O'Farrell, P. J. Halling and M.-C. Parker, Stability of protein-coated microcrystals in organic solvents. *Journal of Molecular Catalysis B: Enzymatic.* **2005**, 33, 65-72.

[72] J. J. Zheng, L. Xu, Y. Liu, X. Y. Zhang and Y. J. Yan, Lipase-coated K2SO4 micro-crystals: preparation, characterization, and application in biodiesel production using various oil feedstocks. *Bioresource Technology.* **2012**, 110, 224.

[73] S. Shah, A. Sharma, M. N. Gupta, Cross-linked protein-coated microcrystals as biocatalysts in non-aqueous solvents. *Biocatalysis and Biotransformation*. **2008**, 26, 266.

[74] M.A. Bradford, A rapid and sensitive method for the quantitation of microgram quantities of protein utilizing the principle of protein-dye binding, *Analytical biochemistry*. **1976**, 72, 248.

[75] S.M. Gotham, P.J. Fryer, W. Paterson, The measurement of insoluble proteins using a modified Bradford assay, *Analytical*

c. Measurement of the concentration of modified lipases using the Bradford method

The free lipases, CCL and PPL, are immobilised on two solid supports, celite and silica. They are then encapsulated with four types of sorbitan ester: sorbitan monostearate (span 60), sorbitan tristearate (span 65), sorbitan monooleate (span 80) and sorbitan trioleate (span 85). The Bradford method is a colorimetric assay that uses a dye, brilliant Coomassie blue G-256, which binds to proteins (our enzyme) by non-covalent interactions at acidic pH. It is based on the change in absorbance (measured at 595 nm), manifested by the change in colour after binding (complexation) with the basic amino acids (arginine, histidine, lysine) and the hydrophobic amino acid residues present in the protein(s).

The unknown concentration of the enzyme is then determined by reading the corresponding absorbance. The equation of the calibration curve is used to calculate its value[81,82].

- Calculation of modified enzyme concentrations

According to the Bradford protein assay method[76] , the unknown protein concentration can be determined using the BSA (*Bovine Serum Albumin*) calibration curve, which gives the following equation: **f(C)= a A + b.**

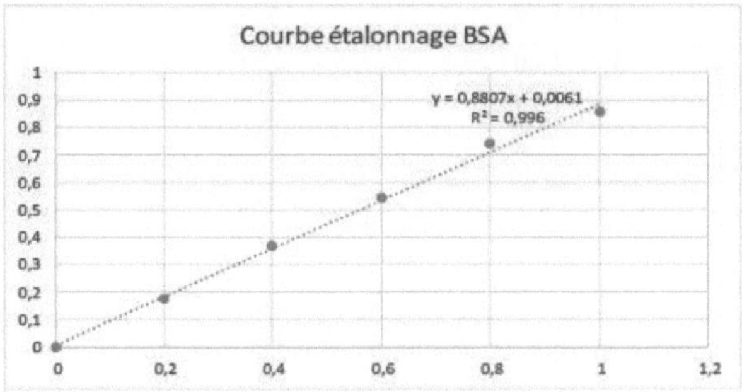

Figure 6- *BSA calibration curve*

Using the relationship derived from the standard curve "Y= 0.8807 X + 0.0061", the concentration of modified lipases is calculated:

Y: absorbance at λ=595nm

X: protein concentration

The results obtained are shown in the tables below.

Table 1- *Influence of the modification in the case of the CDC.*

Lipase	C (mg/ml)	TP (%)

biochemistry. **1988**, 173, 353.

[76] M. M. Bradford, A rapid and sensitive method for the quantitation of microgram quantities of protein utilizing the principle of protein-dye binding. *Analytical biochemistry*, **1976**, *72*, 248.

CCL on	X=0,650	32,5
CCL Is	X=0,565	28,25
CCL C80	X=0,700	35
CCL C85	X=0,500	25
CCL C60	X=0,580	29
CCL C65	X=0,825	41,25

C- *Concentration,* ***TP-*** *Protein content, Rdtim- Immobilisation efficiency,* ***Ic-*** *Immobilisation on celite,* ***Is-*** *Immobilisation on silica,* ***C80-*** *Coted span 80,* ***C85-Coted*** *span 85,* ***C60-*** *Coted span 60,* ***C65-*** *Coted span 65.*

Table 2- *Influence of the modification in the case of the PPL.*

Lipase	C (mg/ml)	TP (%)
PPLIc	X=0.570	28.5
PPL Is	X=0.520	26
PPL C80	X=0.835	41.75
PPL C85	X=0.750	37.5
PPL C60	X=0.780	39
PPL C65	X=0.860	43

C- *Concentration,* ***TP-*** *Protein content, Immobilisation efficiency,* ***Ic-*** *Immobilisation on celite,* ***Is-*** *Immobilisation on silica,* ***C80-*** *Coted span 80,* ***C85-Coted*** *span 85,* ***C60-*** *Coted span 60,* ***C65-*** *Coted span 65.*

Modification of the free enzymes by adsorption on celite gave protein contents of 32.5% for CCL and 28.5% for PPL. With silica we obtain 28% and 26% respectively for CCL and PPL. This shows that celite can be a suitable support for immobilising free lipases .[77]

These materials offer some interesting characteristics. They have pores of uniform diameter, a very high specific surface area and a large pore volume. They are also inert and stable at high temperatures. Lipases have a surface with a hydrophobic character, and hydrophobic interactions probably play an important role in the adsorption process. Physical bonds such as Wan der Waals forces may also come into play and promote contact between the lipase and the support. The technique therefore involves confining the enzyme in a space or network to restrict its mobility. The confinement of enzymes to the surface of macropores (several micrometres in diameter) makes them continuously accessible to reagents in solution. In addition, the nature of the immobilisation support enables water to retained in the medium,

[77] R. Reshmi, S. Sugunan, Superior activities of lipase immobilized on pure and hydrophobic clay supports:Characterization and catalytic activity studies, *Journal of Molecular Catalysis B: Enzymatic* **2013,** 97, 36.

thereby increasing activity of the enzyme .[78]

Modification of the free enzymes by coating gives varying protein contents. The highest values are obtained with span 65, 41% and 43% respectively with CCL and PPL.

The new coated lipases were obtained by aggregation of lipase and surfactant in a buffer solution followed by sonication[79] . The protein concentration was equal in all experiments and dispersed in an aqueous system. The yellow powder obtained is insoluble in water. The effect of the span source on the protein content of the surfactant-lipase complex was investigated. The protein content of the lipase-surfactant complex depends significantly on the source of the span used for coating. It is influenced by the texture of the span. It is also influenced by the length and number of cams of the fatty acid residue of the sorbitan ester. The structural characteristics of the surfactants used are listed the table below.

***Table 3-** Physico-chemical characteristics of spans*

Ester of sorbitati	Features	BHL	TPCCL (%)	TPPPL (%)
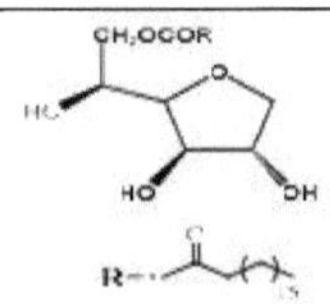	**Span 60** Three hydroxyl groups and a saturated chain.	4,7	29	39
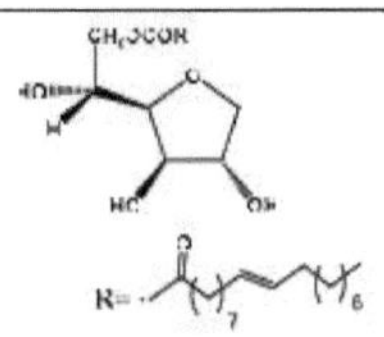	**Span 80** Three hydroxyl groups and an unsaturated chain.	4,3	35	41,75
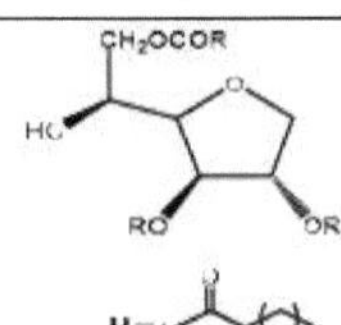	**Span 65** One hydroxyl group and three saturated chains.	2,1	42	43
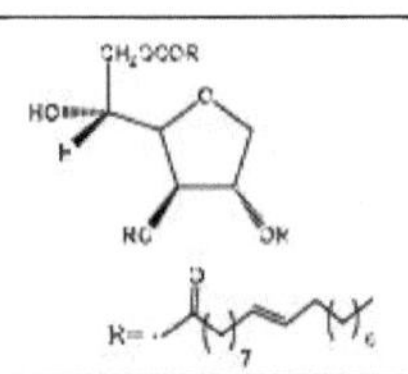	**Span 85** One hydroxyl group and three unsaturated chains.	1,8	25	37,5

[78] L. F. Garcia, V. Gotor, Lipase-catalyzed transesterification in organic media: Solvent effects on equilibrium and individual rate constants. *Biotechnology and bioengineering*.**1998**, 59, 684.
[79] M. Persson, I. Mladenoska, E. Wehtje, P. Adlercreutz, Preparation of lipases for use in organic solvents. *Enzyme and Microbial Technology*, **2002**, *31*, 833.

According to the figures obtained, the protein content (TP) of the resulting complex depends on the hydrophobic-lipophilic balance (HLB) of the surfactant used during the modification process. The results show that higher quantities of the enzyme-surfactant complex are obtained when the lipase is coated with a sorbitan ester containing a three-chambered fatty acid residue of the surfactant. The lower the HLB of the surfactant, the higher its hydrophobicity, so the protein content is higher.

This is probably due to the hydrophobic nature of lipase, the surface of which is mainly made up of non-polar amino acids[80] . Non-polar and weakly polarisable molecules tend to group together to form complexes in which hydrophobic interactions occur between certain protein fractions and the surfactant[81] . Hydrogen bonds between hydroxyl groups and the free amine moieties of the lipase can also play an important role in stabilising the complex .[82]

The immobilisation process preserves fairly large quantities of protein in both the immobilisation and encapsulation processes.

[80] V.V. Mozhaev, I.V. Berezin, K. Martinek. Structurestability relationship in proteins: fundamental tasks and strategy for the development of stabilized enzyme catalysts for biotechnology, *Critical Reviews in Biochemistry*. **1988**, 3, 235.

[81] Okahata Y. and Mori T., Lipid-coated enzymes as efficient catalysts in organic media, *Trends Biotechnology, 1997,* 4, 15.

[82] Y. Okahata, K. Ijiro. A lipid-coated lipase as a new catalyst for triglyceride synthesis in organic solvents, *Journal of the Chemical Society, Chemical Communications*. **1988,** 1, 1392.

III. Structural study of modified *Candida cylindracéa* lipase

1. By encapsulation with span 60 (CCL C60) and 65 (CCL C65)

To characterise *Candida cylindracéa* lipase modified by span 60 (CCL C60) and 65 (CCL C65), we used Scanning Electron Microscopy (SEM) combined with X-ray Dispersive Energy (EDX) microanalysis and X-ray Diffraction (XRD).

A. Scanning Electron Microscopy (SEM) coupled with EDX

Scanning electron microscopy is a technique observing the surface of large biological species on a nanometric scale. Coupled with EDX, it enables us to visualise the external morphology of modified lipases and to determine their chemical composition. At the end of the lipase immobilisation process on the chosen support, confirmation of adsorption is carried out first using SEM and then DRX .[83]

The images obtained for the free CCL, spans 60, 65 and the surfactant-coated biocatalyst are shown in the following figures:

***Figure** 7- SEM microscopic observation of free CCL*

Figure 7 shows the SEM image of an aggregate of the native lipase powder of *Candida cylindracèa*, which is predominantly crystalline.

[83] K. Ramani, S. Karthikeyan, R. Boopathy, L. John Kennedy, A.B. Mandala, G. Sekaran. Surface functionalized mesoporous activated carbon for the immobilization of acidic lipase and their application to hydrolysis of waste cooked oil: Isotherm and kinetic studies. *Process biochemistry*. **2012**, 47, 435.

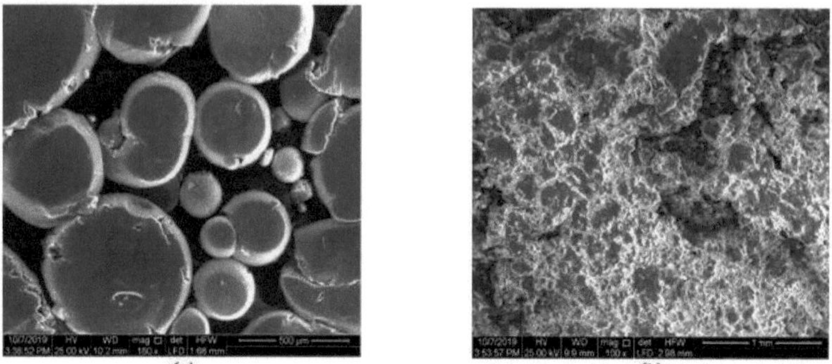

(a) *(b)*

Figure 8- *SEM microscopic observations of* ***(a)*** *span 60 and* ***(b)*** *CCL C60*

Figures 8a and 8b show SEM images of span 60 before and after complexation with lipase, respectively.

It is clear from the SEM image shown in figure 8b that the lipase is well encapsulated by the span[84] . Similar results have been obtained in the literature[85] . The results of EDX microanalysis also suggest that the lipase has been well encapsulated by span 60 (Table 4).

Table 4- *EDX data for CCL C60.*

	CCL		***Span 60***		***CCL C60***	
Element	% of mass	atomic %	% mass	atomic %	% mass	% atomic
CK	10.91	13.71	81.48	85.37	47.14	54.02
NK	17.01	18.32			4.02	3.95
O K	72.08	67.98	18.51	14.63	48.85	42.03

The micrograph clearly shows a difference in morphology which is due to the presence of small shapes adsorbed on the surface of the support. This is further suggested by the results of the EDX analysis carried out on these points, where we obtained mass percentages of the elements carbon, nitrogen and oxygen between those obtained for the support and the biocatalyst separately.

The results of the EDX analyses presented in Table 4 give the elemental compositions of the various elements present in the samples studied. Microanalysis of the native lipase powder

[84] N. Rezgui, C. Bidjou-Haiour. Physical and chemical study of coated *Candida cylindracea* with sorbitan esters. *Research Journal of Chemistry and Environment.* **2021**, 25, 10.

[92] K. Ramani. S. Karthikeyan. R. Boopathy, L. John Kennedy. A.B. Mandala, G. Sekaran. Surface functionalized mesoporous activated carbon for the immobilization of acidic lipase and their application to hydrolysis of waste cooked oil: Isotherm and kinetic studies", *Process biochemistry.* **2012,** 47, 435.

revealed three specific majority peaks for carbon, nitrogen and oxygen: 10.91%, 17.01% and 72.08% respectively.

- The peaks for span 60 are carbon (81.48%) and oxygen (18.51%).
- For the modified CCL C60 lipase, the peaks for carbon (47.14%), nitrogen

(4.02%) and oxygen (48.85%) (Table 4).

The modified lipase composition clearly shows the presence of a nitrogen peak that does not exist in the carrier composition, confirming encapsulation.

The lipase is also coated with span 65. Figures 9a and 9b show SEM images of span 65 before and after complexation with lipase, respectively.

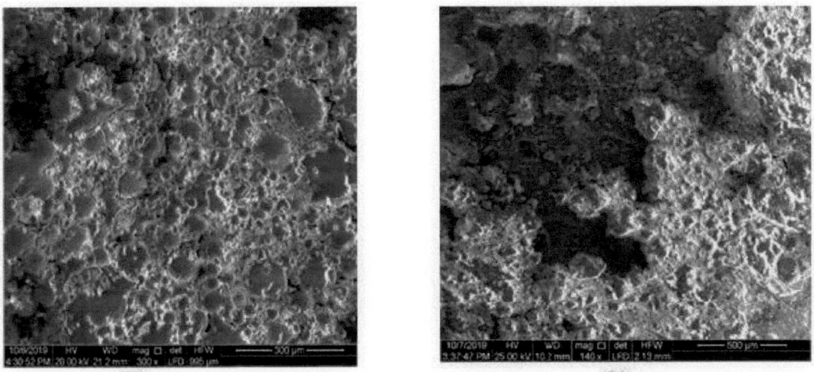

a) b)

Figure 9- *SEM microscopic observations of **a)** span 65 and **b)** CCL C65*

The same observations were made for CCL C65, namely the good integration of the lipase on its support, confirmed by the SEM images.

The EDX microanalysis results also suggest that the lipase was well coated by span 65 (Table 5).

Table 5- *EDX data for CCL C65.*

	CCL		***Span 65***		***CCL C65***	
Element	% of mass	atomic %	% of mass	atomic %	% of mass	atomic %
CK	10.91	13.71	83.98	86.21	48.82	55.53
NK	17.01	18.32			6.22	6.07
O K	72.08	67.98	16.01	13.79	44.97	38.40

This table shows the elemental composition in percentages of the powders obtained by EDX. These results clearly show the presence of oxygen, carbon and nitrogen. As can be seen, microanalysis of the native lipase powder shows three specific majority peaks of carbon, nitrogen and oxygen respectively 10.91%, 17.01% 72.08%.

- The peaks for span 65 are carbon (83.98%) and oxygen (16.01%).
- For the modified CCL C65 lipase, the values for these same atoms are (48.82%) and (44.97%) respectively (Table 5).

Mass percentages of the elements carbon, nitrogen and oxygen between those obtained for the support and the biocatalyst separately were observed, with the appearance of the nitrogen peak (6.22%) for the coated biocatalyst. This confirms that the free CCL was successfully encapsulated by span 65.

B. X-ray Diffraction (XRD)

X-ray diffraction enabled us to determine the location of the adsorbed enzyme[86] . The diffractograms obtained for the different enzymes synthesised and for the free lipase are shown in Figures 10 and 11. The XRD spectra are recorded in RAW files. They show that sorbitan esters are successfully integrated into the lipase structure.

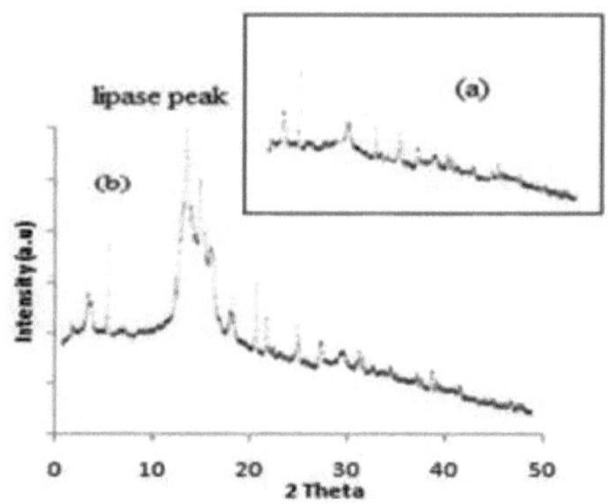

Figure 10- *XRD spectra of* ***(a)*** *span 65 and* ***(b)*** *CCL C65*

[86] S. Gupta., S. Yogesh, S. Javiya, M. Bhambi, C.S. Pundir, K. Singh and A. Bhattacharya. Comparative study of performances of lipase immobilized asymmetric polysulfone and polyether sulfone membranes in olive oil hydrolysis. *International journal of biological macromolecules.* **2008,** 42, 145.

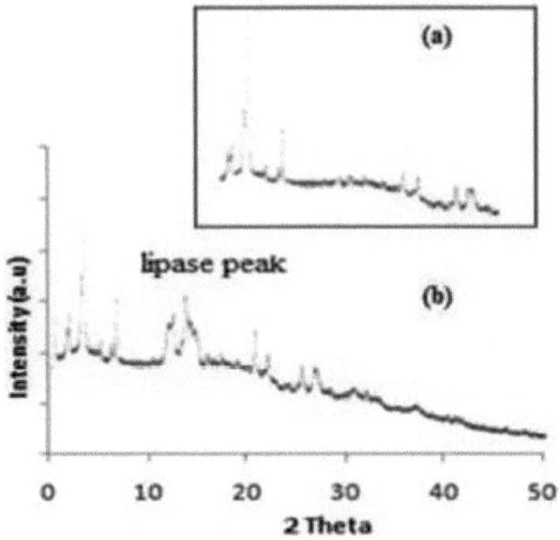

__Figure 11-__ XRD spectra of __(a)__ span 60 and __(b)__ CCL C60.

From the above figures, the coating of lipase by sorbitan esters is evidenced by the additional lipase peak in Figures 10 and 11. These figures also show the intensity of the lipase peak to be higher for adsorption on span 65 (Figure 10), is probably due to a higher amount of lipase. The protein content of the lipase-surfactant complex depends significantly on the source of the span. It is also influenced by the length and chain number of the fatty acid residue of the sorbitan ester, and therefore the hydrophilic-lipophilic balance (HLB) of the surfactant as explained above.

It can also be seen that the binding of lipase to the supports does not affect the structure of the phases that make up spans 60 and 65. No change was observed in the characteristic peaks except for the appearance of the additional lipase peak.

2. By immobilisation on celite (CCL Ic)

In order to prove the adsorption of *Candida cylindracéa* lipase on celite (CCL Ic), we used the Scanning Electron Microscopy (SEM) technique combined with X-ray Dispersive Energy (EDX) microanalysis.

The images obtained for the solid support and the lipase adsorbed on the celite are shown in the following figures:

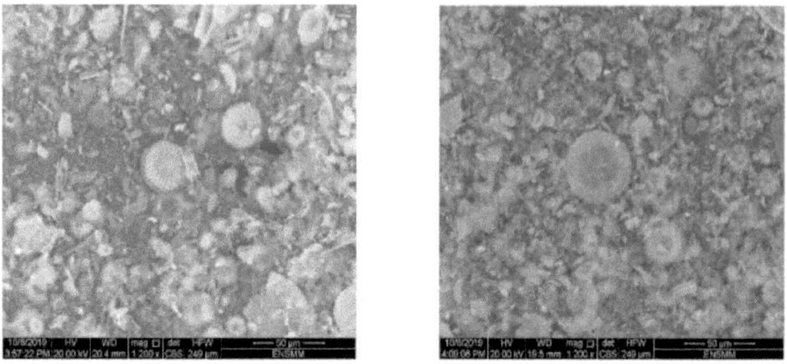

(a) (b)

Figure 12- *SEM microscopic observations of* ***(a)*** *celite and* ***(b)*** *CCL Ic*

Comparison of the SEM image obtained for the free lipase (Figure 12) and the images in figures a for the blank support and b for the support loaded with lipase clearly shows the adsorption of the biocatalyst onto the celite. This is also confirmed by the results of the EDX analysis (Table 6).

Table 6 - *Data from EDX analysis of CCL.*

	CCL		***Celite***		***CCLIc***	
Element	% mass	atomic %	% mass	atomic %	% mass	atomic %
CK	10.91	13.71	10.78	13.54	6.83	8.80
NK	17.01	18.32	17.17	18.50	8.65	9.54
O K	72.08	67.98	72.05	67.96	84.52	81.66

The EDX results presented in Table 6 are obtained by XRD processing, which enables the energy (KeV) of the lines be analysed in order to identify the atoms present in the zone pointed at by the probe and to obtain the elemental atomic compositions of the different powders. These results clearly show the presence of oxygen, carbon and nitrogen.

The dispersive spectrum of the native lipase powder shows three intense and specific majority peaks of carbon, nitrogen and oxygen respectively 10.91%, 17.01% 72.08%.

For celite the peaks are : C (10.78%), N (17.17%) and O (72.05%). For the modified lipase CCL Ic, the values for these same atoms are respectively: (6.83%), (8.65%) and (84.5%). The mass percentages of carbon, nitrogen and oxygen elements between those obtained for the support and the biocatalyst separately show the successful immobilisation of the lipase on its support.

V. Structural study of modified *porcine pancreatic* lipase

a. By encapsulation with span 65 (PPL C65)

To characterise Span 65-modified *porcine pancreatic* lipase (PPL C65), we used Scanning Electron Microscopy (SEM) combined with X-ray Dispersive Energy (EDX) microanalysis. The images obtained for free PPL, span 65 and surfactant-coated biocatalyst are shown in the following figures:

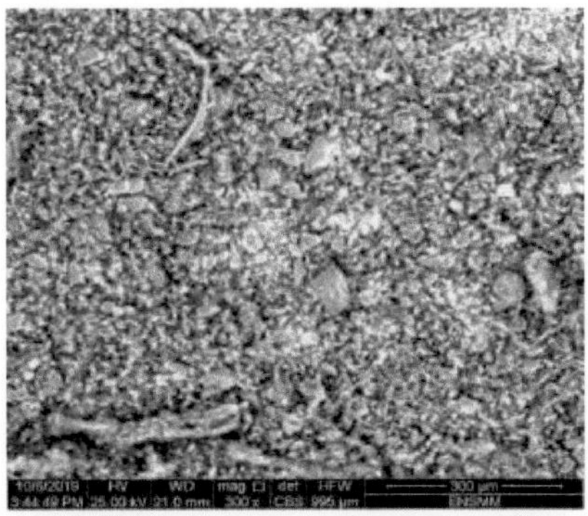

***Figure** 12-SEM microscopic observation of free PPL*

Figure 12 shows the SEM image of an aggregate of *porcine pancreatic* lipase powder which is predominantly crystalline.

(a) *(b)*

***Figure** 13-SEM microscopic observations **of (a)** span 65 and **(b)** PPL C65*

Figures 13a and 13b show SEM images of virgin span 65 and lipase after coating, respectively.

Scanning electron microscopy coupled with EDX enabled us to visualise the external morphology of the virgin celite and the celite loaded with PPL, in order to confirm the adsorption of the lipase onto the support.

The results of the EDX microanalysis confirm that the lipase was indeed coated with span 65

(Table 7).

***Table** 7- EDX analysis data for PPL C65.*

	PPL		***Span 65***		***PPL C65***	
Element	% mass	atomic %	% mass	atomic %	% mass	atomic %
CK	17.60	20.97	21.93	25.16	19.23	22.63
NK	42.18	43.08			47.91	48.34
O K	40.21	35.96	78.06	74.84	32.86	29.03

The EDX analysis presented in Table 7 made it possible identify the atoms present in the pointed zone and to obtain the elemental atomic compositions of the different powders. These results clearly show the presence of oxygen, carbon and nitrogen.

The dispersive spectrum of the native lipase powder shows three specific majority peaks of carbon, nitrogen and oxygen respectively 17.60%, 42.18% and 40.21%.

For span 65, the peaks were carbon (21.93%) and oxygen (78.06%). For the modified lipase CCL Ic, the values for these same atoms are respectively: (19.23%), (47.91%) and (32.86%).

The appearance of the nitrogen peak for the modified lipase and the mass percentages of carbon, nitrogen and oxygen elements included or close to those obtained for the support and biocatalyst separately, show the successful immobilisation of the lipase on its support.

b. By immobilisation on celite (PPL Ic)

In order to prove the adsorption of *porcine pancreatic* lipase on celite (CCL Ic), we used the technique of Scanning Electron Microscopy (SEM) combined with X-ray Dispersive Energy (EDX) microanalysis.

The images obtained for the solid support and the lipase adsorbed on the celite are shown in the following figures:

(a)

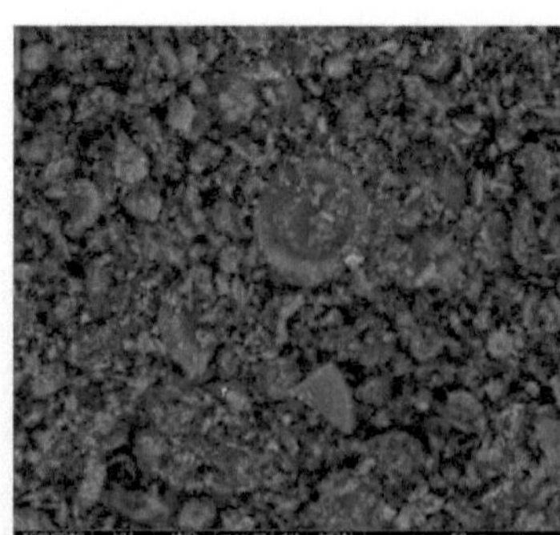

(b)

Figure 14- *Microscopic observations* ***(a)*** *celite,* ***(b)*** *PPL Ic by SEM*

Table *8- EDX analysis data for PPL Ic.*

	PPL		***Celite***		***PPL Ic***	
Element	% mass	atomic %	% mass	atomic %	% of mass	atomic %
CK	17.60	20.97	10.78	13.54	1.43	1.63
NK	42.18	43.08	17.17	18.50	98.57	98.37
O K	40.21	35.96	72.05	67.96		

The EDX analysis presented in Table 8 made it possible identify the atoms present in the pointed zone and to obtain the elemental atomic compositions of the different powders. These results clearly show the presence of oxygen, carbon and nitrogen.

The dispersive spectrum of the native lipase powder shows three specific majority peaks for carbon, nitrogen and oxygen: 17.60%, 42.18% and 40.21% respectively.

For celite the peaks are : C (10.78%), N (17.17%) and O (72.05%).

For the modified lipase PPL Ic, the values for the carbon and nitrogen atoms are (101.43%) and (98.57%) respectively.

The mass percentages of carbon, nitrogen and oxygen elements, at or close to those obtained for the support and biocatalyst separately, show the success of the immobilisation of the lipase on its support.

CHAPTER 4

EDX spectra

a. EDX spectra of free lipases

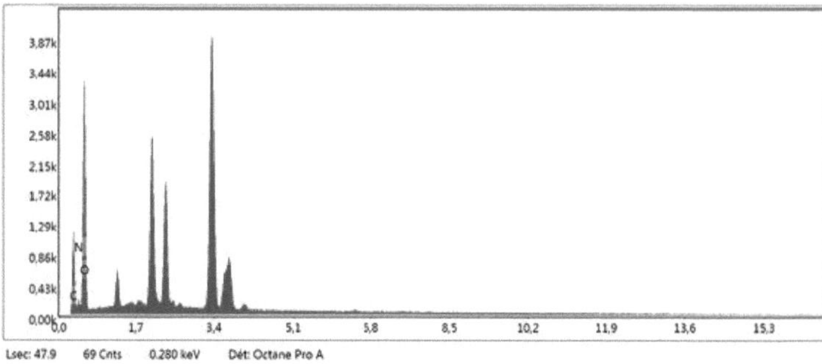

Figure 11- *EDX microanalysis electron spectrum of CCL*

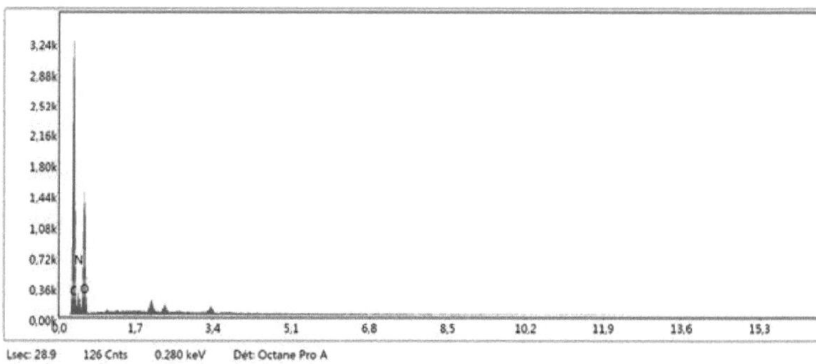

Figure 12- *EDX microanalysis electron spectrum of PPL*

1. EDX spectra of modified enzymes

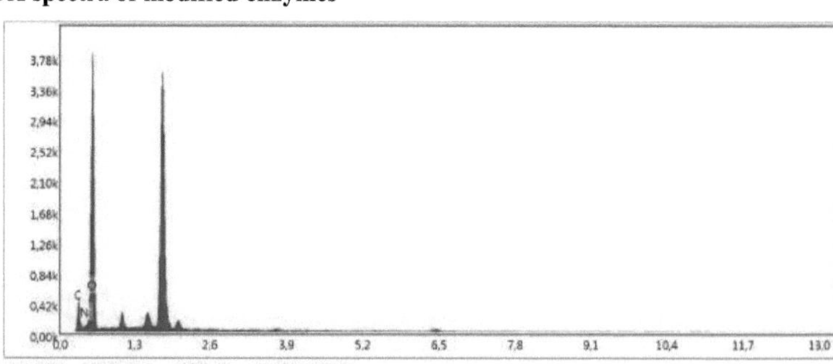

Figure 13- *EDX microanalysis electron spectrum of CCL Ic*

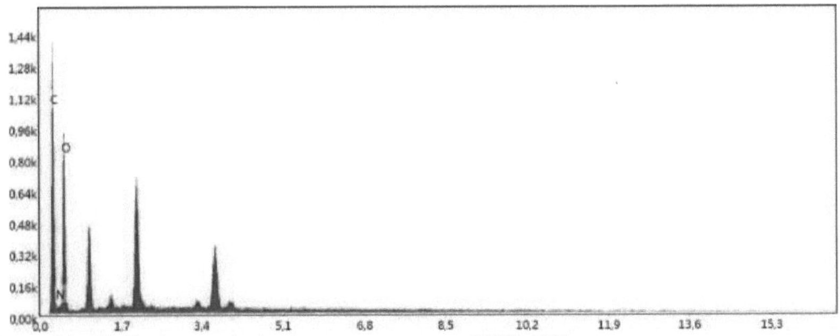

Figure 14- *EDX microanalysis electron spectrum of CCL C65*

Lsec: 28.9 O Cnts 0.000 keV Det: Octane Pro A

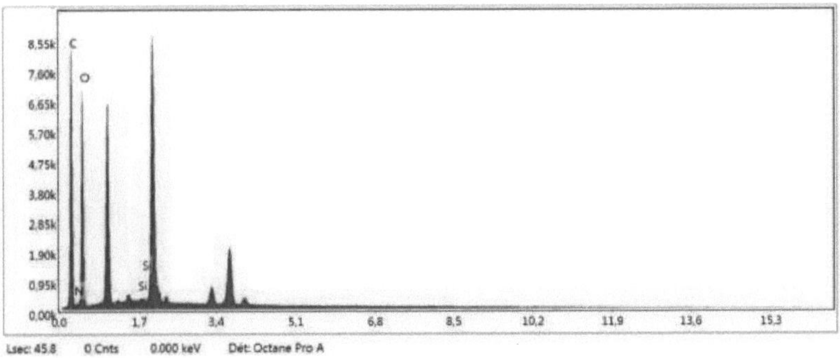

Figure 15- *EDX microanalysis electron spectrum of CCL C65*

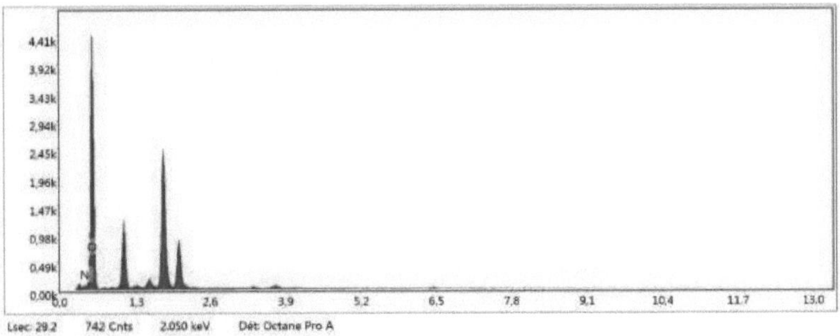

Figure 16- *EDX microanalysis electron spectrum of PPL Ic.*

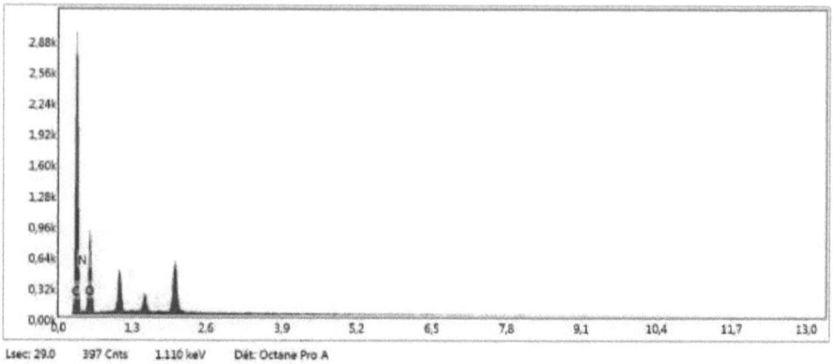

***Figure 17**- EDX microanalysis electron spectrum of PPL C65*

CHAPTER 5

Effect of lipase modification in the enzymatic esterification reaction of D-glucose

To assess the effectiveness of the modified enzymes, they were also tested in the enzymatic esterification reaction of a-(+)-D-glucose under the operating conditions defined previously.

Reaction

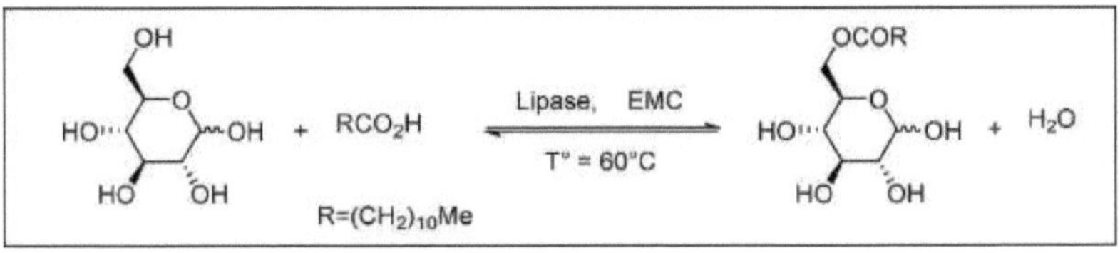

Diagram 2- *Enzymatic esterification reaction of D-glucose*

The progress of the esterification reaction is based on the nature of the lipase used as biocatalyst. The enzymatic activity of the different lipases modified on a laboratory scale was compared first with that of free lipase and then with that of one-pot lipase. The reactions took place in the EMC at 60°C, with 30 mg lipase and 30 mg molecular sieve.

The results are shown in the tables below.

a. The case of the CDC

The kinetics obtained with the different forms of CCL are illustrated in the table below.

Table 9 - *Influence of lipase on conversion in the case of CCL.*

Enzyme	Conversion			
	2h	**24h**	**48h**	**72h**
CC1 free	3	8	8	14
CCL Ic	0	11	19	45
CCL Is	0	17,5	21	21,50
CCL C80	18	17	10	7
CCL C85	5	20	24,50	22
CCL C60	12,50	28,50	31,70	31,70
CCL C65	10,50	31,70	36,50	50
CCL + spanSO	0	10	12	15
CCL + span85	3	10	11	12
CCL + span60	5	12	12	18
CCL + span65	5	15	14	15

Reaction conditions: *1 mmol a-(+)-D-glucose, 1eq Al, 5 ml EMC, 30mg TM, T°=60°C.*

According to the results presented in Table 9, and comparing the conversions obtained for free and modified CCL under the same conditions, it can be seen that conversion reached a rate of 45% for CCL immobilised on celite (CCL Ic) and 50% for CCL coated with span 65 (CCL C65). Conversion with the free CCL enzyme was only 14%. Both methods are therefore effective and significantly increase the enzymatic activity of CCL.

The reactions were also carried out in the presence of lipase and surfactant in a single pot and the conversions obtained did not exceed 18%. This is further evidence of the effectiveness of the modified CCL.

b. PPL

The kinetics obtained with the different forms of PPL are shown in Table 10.

Table 10 - Influence *of lipase on conversion in the case of PPL.*

Enzyme	Conversion			
	2h	24h	48h	72h
Free PPL	0	0	10	12
PPL Ic	0	22	20	20
PPL Is	0	18	20	20
PPL C80	10	10	28	48
PPL C85	15	18,2	30	42
PPL C60	5	23	38,5	44,5
PPL C65	10	26,70	36	50
PPL + spanSO	5	8	10	12
PPL + span85	0	0	15	15
PPL + spanôO	0	5	5	5
PPL + span65	5	8	8	8

Reaction conditions*: 1 mmol a-(+)-D-glucose, 1eq Al, 5 ml EMC, 30mg TM, T°=60°C.*

In the case of PPL, immobilisation slightly increases conversion compared with free lipase. Conversion is 20% whatever the immobilisation support (silica or celite).

Coating the PPL with spans resulted improved enzymatic activity compared with immobilisation, with significant conversions, the highest being obtained for PPL coated in span 65 (50%).

The addition of spans (80, 85, 60, 65) to the free enzyme in its reaction medium had no effect on its activity. The conversion values obtained in the presence of spans in one pot with lipase show the importance of the coating. The encapsulated enzyme becomes stable and active in an organic medium.

The results obtained prove that coating is the best technique for enhancing catalytic activity while preserving the enzyme structure. The protein content obtained by this method remains the highest compared with immobilisation.

Catalysis of the D-glucose esterification reaction by the modified biocatalysts under the same operating conditions therefore indicates that the coated enzymes have the highest catalytic activity, with a conversion of 50% for CCL and PPL coated with span 65.

c. Comparison of modified lipases with CAL B

The enzymatic activities of CCL and PPL coated with span 65 were then compared with those of *Candida antractica* B lipase, a commercial lipase[102] , immobilised on resin and conventionally used in the synthesis of sugar esters.

Table 11 *- Comparison of modified lipases with CAL B.*

Enzymes	Conversion			
	2h	24h	48h	72h
CAL B	10	44,5	50,5	52,5
PPL C65	10	26,70	36	50
CCL C65	10,50	31,70	36,50	50

Reaction conditions*: 1 mmol a-(+)-D-glucose, 1eq Al, 5 ml EMC, 30mg TM, T°=60°C.*

It can be seen that after 72h of reaction at 60°C, 52.50% conversion was obtained during the enzymatic acylation reaction of glucose with CAL B. Comparison with the best results obtained with modified lipases in the same reaction showed activities similar to those observed with *Candida antartica* B lipase (PPL C65, 50%, CCL C65, 50%) however, after 72h.

The modified enzymes obtained in this therefore have a catalytic activity comparable to that of CAL B, which is the lipase most widely used for the enzymatic synthesis of sugar esters. These lipases show high catalytic activity and thermal stability compared with the native enzyme, whose operating temperature does not exceed 40°C [103104105106] . In addition, a protection system has been developed to minimise direct contact of the organic solvent and substrates with the lipase in order to facilitate the lipolytic esterification of sugars, the aim

[102 94] R. BenSalah, H. Ghamghui, N. Miled, H. Mejdoub, Y. Gargouri, Production of butyl acetate ester by lipase from novel strain of Rhizopus oryzae. *Journal of bioscience and bioengineering*. **2007**, *103*, 368.
[95] a. P. Degn, W. Zimmerma, Optimization of carbohydrate fatty acid ester synthesis in organic media by a lipase from Candida antarctica. *Biotechnology and* **bioengineering2001**, *74*, 483.
b. Y. Li, D. Rethwisch, Scale-up of pseudo solid-phase enzymatic synthesis of α-methyl glucoside acrylate. *Biotechnology and bioengineerring*,. **2002**,*79*, 15.

[103] P. Adlercreutz. Immobilisation and application of lipases in organic media. *Chemical Society Reviews*, **2013**, *42*, 6406.
[104] M. K.Walsh, R. A Bombyk, A. Wagh, A. Bingham, L. M. Berreau, Synthesis of lactose monolaurate as influenced by various lipases and solvents. *Journal of Molecular Catalysis B: Enzymatic*, **2009**, 60, 171.
[105] T. Kobayashi, Lipase-catalyzed syntheses of sugar esters in non-aqueousmedia, *Biotechnology. Letters*. **2011,** 33, 1911.
[106] Z. Yang, Z.-L. Huang, Enzymatic synthesis of sugar fatty acid esters in ionic liquids, *Catalysis Science & Technology*. **2012,** 2 1767.

being to make these biocatalysts (PPL and CCL) more attractive from a synthetic point of view. The high catalytic activities obtained are probably due to better dispersion of the catalyst in organic media.

VIII. Solvent effect in the enzymatic esterification reaction of D-glucose

A suitable solvent is essential for lipase-catalysed synthesis of sugar esters [97,98,99]. The solvent must be able to dissolve sufficient quantities of both substrates during the enzymatic esterification reaction. The problem associated with this study is therefore to find a compromise so that the solvent solubilises both the sugar (polar) and the fatty acid (apolar). In addition, it must not compromise the stability of the enzyme or its activity.

The PPL encapsulated with span 65 (PPL C65) is subjected to extreme solvent conditions in the enzymatic esterification reaction of D-glucose. Organic solvents with different polarities were tested. Ionic liquids were then examined as a replacement for organic solvents[107] . They represent a new class of solvents offering an opportunity for cleaner chemistry and sustainable development.

The addition of a co-solvent can also help to improve solubility[108] . Consequently, we chose dimethyl sulphoxide, DMSO, which is a very polar solvent and should completely dissolve D-glucose.

a. Influence of solubility on the esterification reaction

1. Solubility test

In order to understand the solubility of the sugar and its importance in the synthesis, different types of solvents with different polarities were tested[109] . These included organic solvents: tetrahydrofuran (THF), 2-methylbutan-2-ol (2M2B), dimethylsulphoxide (DMSO) and ethylmethylketone (EMC); imidazolium ionic liquids: 1-butyl-3-methyl-imidazolium tetrafluoroborate (LI1) and 1-ethyl-3-methyl-imidazolium tetrafluoroborate (LI2),

Table 12 shows the solubility of the solvents used in the enzymatic esterification of D-glucose.

Table 12- *Solubility of D-glucose.*

Solvents	LogP (pH=7, 20°C) (ECHA)	Solubility (100mg/2ml)
DMSO	-1,35	Ires good
EMC	0,29	Good

[107] L. Xiao-Sheng, W. Qing, H. Ze-Lin, C. Yu-Zheng, J. Peter. Z. Hallingc, Impacts of ionic liquids on enzymatic synthesis of glucose laurate and optimization with superior productivity by response surface methodology. *Process Biochemistry*, **2015,** 50, 1852.

[108] A. Ghanem, Trends in lipase-catalyzed asymmetric access to enantiomerically pure/enriched compounds. *Tetrahedron*, **2007**, 63, 1721-54.

[109] L. Xiao-Sheng, W. Qing, H. Ze-Lin, C. Yu-Zheng, J. Peter. Z. Hallingc, Impacts of ionic liquids on enzymatic synthesis of glucose laurate and optimization with superior productivity by response surface methodology. *Process Biochemistry*, **2015,** 50, 1852.

THE	0,52	Good
2M2B	0,89	Wrong
Lil	4,13	Wrong
LI2	5,06	Wrong

***ECHA**: European Chemicals Agency.*

The results of the D-glucose solubility test show that the best solvent for almost complete dissolution of the sugar is DMSO. The other organic solvents dissolve it more or less well (Table 12).

The solubility of sugar in ionic liquids is not good, however these solvents are desirable in enzyme synthesis as they generally do not affect the catalytic activity of the enzyme.

b. Influence of the reaction medium

The reactions were first carried out in organic solvent medium alone, then in ionic liquid medium. In order to further optimise sugar conversion, an appropriate quantity of DMSO was introduced into the reaction medium. A kinetic study is carried out with different amounts of this polar solvent: (solvent/DMSO): (lml/IOμI, 1ml/20gl, 1ml/30gl, lmb'4-θμΐ, 1ml/50gl, 1ml/60gl, 1ml/70gl, 1ml/80gl, 1ml/90gl, lml/IOOμI).

Preliminary results from this study show that the best ratio is (v/v): (1ml/20gl) and (^1/50μ1). This is confirmed by the literature[110] . The conversions obtained with the other ratios studied are not conclusive and are therefore not reported.

The results are shown the tables below.

- **Reactions in solvent medium alone**

The results obtained for the progress of the enzymatic esterification reaction of glucose in the presence of PPL C65 with the different solvents are shown the table below.

***Table 13** - Influence of solvent on conversion in the case of PPL C65.*

Solvent	Conversion			
	2h	24h	48h	72h
EMC	10	26,70	36	50
THE	10	20	32	45
2M2B	10	18	30	40,5
ILS1	5	28	34	43,5
ILS2	8	22	32	42

***Reaction conditions**:* 1 mmol *a-(+)-D-glucose, 1 eq Al, 30 mg PPL C65, 30mg TM, T°=60°C.*

In organic media, PPL C65 gave a maximum conversion in EMC of 50%. In molten salts with

110 X.Sh. Lina, Q. Wenb, Z.L. Huanga, C. Yu-Zheng, J. Peter. Impacts of ionic liquids on enzymatic synthesis of glucose laurate andoptimization with superior productivity by response surfacemethodology. *Process Biochemistry.* **2015**, 50, 1852.

the same lipase, the conversions obtained were of the order of 40% for all solvents.

- **Influence of co-solvent**

In order to improve the solubility of the starting sugar, different portions of DMSO are introduced into the reaction medium.

In comparison with the solvents alone and by adding DMSO with a ratio of (1ml / 50µl) (solvent/DMSO), the conversions obtained for LIs are 54.5% and 50.5% for LI1 and LI2 respectively.

Table 31- *Influence of co-solvent on conversion in the case of PPL C65.*

Solvent	Conversion			
	2h	**24h**	**4811**	**72h**
EMC/ DMSO (1 mi: 50µ1)	15	20	38	51,5
EMC/ DMSO (1 ml: 20µ1)	5	18	25	41,2
THE/ DMSO (1 ml: 50µ1)	10	22	22	45,5
THE/ DMSO (1 ml: 20µ1)	8	19	28	42
2M2B/DMSO (1 ml: 20µ1)	8	10	20	35
2M2B/DMSO (1 ml: 50µ1)	5	20	28	40,5
Lil/ DMSO (1 ml: 20µ1)	18	30	40	48
Lil/ DMSO (1 ml: 50µ1)	20	35	45,5	54,5
LI2/ DMSO (1 ml: 20µ1)	10	26	36	43
LI2/ DMSO (1 ml: 50µ1)	15	29	40,5	50,5

Reaction conditions: 1 mmol *a-(+)-D-glucose, 1 eq Al, 30 mg PPL C65, 30mg TM, T°=60°C.*

No significant changes in conversion were observed during enzymatic esterification reaction in the other solvents by adding the same co-solvent.

The catalytically active 3D structure of enzymes depends on numerous hydrogen bonds, hydrophobic effects, Van der Waals forces and dipolar interactions, which occur within the protein molecule and between the protein and the solvent. So the prerequisites for homogeneous non-aqueous enzyme catalysis are the solubility and stability of the biocatalysts in the organic solvent used.

According to the tables above, the results obtained with LI1 and EMC reproducible. A variation in conversion depending on the nature of the medium and the lipase used was observed.

- The best conversions were in organic solvents alone, i.e. 50% PPL C65 in EMC after 72h.
- In the LIs, the best conversion obtained was 43.5% with PPL C65.

A comparison of the results obtained in the different cases studied shows that the reaction medium has a significant influence on conversion.

The conversions obtained with organic solvents depend on the nature of the solvent (polarity, denaturation capacity) [111112113] . The status of the enzyme (modified or not) affects conversion to the product of lipase-catalysed esterification; conversions to esters are increased when reactions are carried out in moderately polar solvents. Polar solvents have a significant effect on the solubility of substrates, but they can deform the lipase's constitutive aqueous layer, eventually causing denaturation of the enzyme[105,106].

In an organic medium, the solvent, regardless of its polarity, can restrict the conformational flexibility of the enzyme by modifying the water layer necessary for its activity, thereby denaturing its microstructure. This can lead to reversible or irreversible denaturation of the lipase[114] . It has also been found that lipases behave differently in different organic media with different levels of resistance in different reaction systems .[115]

In the interests of green chemistry and optimise the functioning of the modified lipases, LIs are used instead of organic solvents.

We first observed that replacing the organic solvents with ionic liquids did not have a significant influence on the progress of the reaction with all the lipases. The conversions obtained are probably due to the low solubility of the sugar and long-chain acyl substrates in the imidazolium ionic liquids (Solubility table), leading to a two-phase system that reduces the availability of substrates for the enzyme and thus reduces biocatalysed acylation.

Beyond these results, we thought that a better conversion would be obtained if the substrates were completely soluble in the medium. To this end, we introduced a minimal quantity of a polar solvent, DMSO, ensure the solubility of the sugar and the proper functioning of the lipase. A kinetic study was carried out assess the best solvent/DMSO ratio to use.

Initially a screening of DMSO quantity is performed. (solvent/DMSO): (lml/IOμI, 1ml/20gl, 1ml/30gl, 1ml/40gl, 1ml/50gl, 1ml/60gl, 1ml/70gl, 1ml/80gl, 1ml/90gl, lml/IOOμI). However only the proportions (1ml/20μl, 1ml/50μl) provide good conversions.

The addition of DMSO with (solvent/DMSO) portions: (1ml/20μl, 1ml/50μl) significantly increases the conversion, in organic solvents. The best conversion obtained is 51.5% in

---

[111] S. Soultani, J-M. Engasser, M. Ghoul. Effect of acyl donor chain length and sugar/acyl donor molar ratio on enzymatic synthesis of fatty acid fructose esters. *Journal of Molecular Catalysis B: Enzymatic.* **2001**, 11, 31.

[112] S. Hari Krishna, N. G. Karanth, Lipases and lipase-catalyzed esterification reactions in nonaqueous media. *Catalysis Reviews.* **2002,** *44*, 499.

[113] M. M. Soumanou, U. T. Bornscheuer. Improvement in lipase-catalyzed synthesis of fatty acid methyl esters from sunflower oil. *Enzyme and microbial technology*, **2003**, *33*, 97.

[114] S. Torres, G. R. Castro. Non-aqueous biocatalysis in homogeneous solvent systems. *Food Technol Biotechnol*, **2004,** *42*, 271.

[115] J. Yu, J. Zhang, A. Zhao, X. Ma, Study of glucose ester synthesis by immobilized lipase from Candida sp. *Catalysis Communications*, **2008**,*9*,1369.

(EMC/DMSO): (^í/50µí) and 54.5% in (LI1/DMSO): (^í/50µí) with PPL C65.

When comparing LIs in the presence of the same lipase, the high viscosity of the reaction medium may increase the mass transfer limitation, as observed for the benzoylation of floxuridine catalysed by Pseudomonas cepacia lipase (PSL-C) in THF/[Bmim][PF6] where 98% conversion was achieved in 82 h with 25% [Bmim][PF6] while 5% [Bmim][PF6] achieved over 99% conversion after 48.5 h .[116]

This is true for LI1 (C= 54.5%) which is less viscous and therefore more hydrophobic than LI2 (C=50.5%). Lipase is less affected in the first case. In terms activity and stability, the enzymes may prefer hydrophobic LIs to hydrophilic LIs. This is because the hydrophobic medium can protect the essential water layer around the protein molecules, which increases the enzyme's stability in the denaturing environment and reduces direct protein-ion interactions[117] . Protein conformation will also be maintained active as hydrophobic LIs build an ionic matrix to preserve enzyme molecules in tolerable microenvironments .[118]

---

[116] N. Li, D. Ma, M. H. Zong. Enhancing the activity and regioselectivity of lipases for 3'-benzoylation of floxuridine and its analogs by using ionic liquid-containing systems. *Journal of biotechnology*, **2008**, *133*, 103.

[117] A.J. Carmichael, K.R. Seddon, Polarity study of some 1-alkyl-3-methylimidazolium ambient-temperature ionic liquids with the solvatochromic dye, Nile Red. *Journal of Physical Organic Chemistry*, **2000**, 13, 591.

[118] L. Cao, U.T. Bornscheuer, R.D Schmid. Lipase-catalyzed solid-phase synthesis of sugar esters. Influence of immobilization on productivity and stability of the enzyme. *Journal of Molecular Catalysis B: Enzymatic.* **1999**, 6, 85.

Conclusion

New lipases synthesised and characterised. The techniques used to obtain modified lipases from free lipases immobilisation on a solid support and encapsulation by non-ionic surfactants.

The synthesised lipases were characterised using Scanning Electron Microscopy (SEM) combined with X-ray Dispersive Energy (EDX) microanalysis and X-ray Diffraction (XRD). The images obtained for the different forms of lipase were used to characterise the new biocatalysts and to prove the integration of the lipases on the chosen supports.

Lipase from *Candida cylindracéa* immobilised on celite expressed an activity three times greater than that of free lipase. It has been shown that CCL has a surface with very high hydrophobicity[119] and hydrophobic interactions probably play an important role in the adsorption process. This enzyme seems to adsorb easily onto the porous support in order to limit its mobility, which makes it accessible to reagents in solution. Celite is therefore a good adsorption support for CCL. Immobilised *porcine pancreatic* lipase has a low catalytic activity, so desorption must probably take place at some point.

Encapsulation or coating of lipases with non-ionic surfactants is a technique that appears to work better than adsorption. Spans are insoluble in aqueous media, allowing the lipase to be integrated onto the surfactant, which acts as a support material.

Span-modified *porcine pancreatic* lipase has an activity four times greater than that of free lipase. *Candida cylindracéa* lipase encapsulated with spans 80 and 85 has a low catalytic activity. However, its activity is improved by encapsulation with spans 60 and 65. It is multiplied by four in the case of span 65.

Comparison of the catalytic activity of the modified enzymes with CAL B[120] , a commercially available immobilised enzyme, shows that the modified lipases exhibit catalytic activities comparable with those obtained in the presence of CAL B (PPL C65=50%, CCL C65=50%). These results consistently show that modified lipases work better than free lipases. In addition, modification has enabled us increase the thermal stability of these enzymes, whose optimum operating temperature is around 40°C. Beyond these properties, there are environmental compatibilities, with the use of biocatalysts being considered as green chemistry.

---

[119] Y. Sugimura, K. Fukunaga, T. Matsuno, K. Nakao, M. Goto, F. Nakashio. A study on the surface hydrophobicity of lipases, *Biochemical Engineering Journal,* **2000**, 5, 123-128.

[120] D. Coulon, Girardin M.,Rovel B.,Ghoul M., Comparison of direct esterification and transesterification of fructose by *Candida* antarticalipase, *Biotechnology Letters*, **1995,** 17, 2, 183-186.

References

1. E.N. Vulfson, P.J. Halling, H.L. Holland, Enzymes in nonaqueous solvents: methods and protocols, *Humana Press*, **2001**, 701.
2. H. Oka, K. Aimono, S. Tsuyutani, T. Fujita, K. Hashiba. *Eur. Patent EP 492397, Kao Corp*. **1992**.
3. H. Lüders, *Eur. Patent EP 252241, Hüls AG*. **1987**.
4. H.A. El-Baz, A.M. Elazzazy, T.S. Saleh, M. Dourou, J.A. Mahyoub, M.N. Baeshen, H.R. Madian, and G. Aggelis. Enzymatic Synthesis of Glucose Fatty Acid Esters Using SCOs as Acyl Group-Donors and Their Biological Activities. *Applied Sciences*. **2021**, 11, 2700.
5. N. Nicolas, B. Thierry, P. Daniel. Surfactants based on renewable substances. *L'actualité chimique*. **2000**. 70, 75.
6. T. Raku, M. Kitagawa, H. Shimakawa, Y. Tokiwa. Enzymatic synthesis of hydrophilic undecylenic acid sugat esters and thier biodegradability. *Biotechnology letters*. **2003,** 161, 166.
7. T. Raku, M. Kitagawa, H. Shimakawa, Y. Tokiwa. Enzymatic synthesis of trehalose esters having liphophilicity. *Journal of biotechnology*. **2003,** 100, 208.
8. N. do Amaral Sampaio Neta, J.C. Sousa dos Santos, S. de Oliveira Sancho, S. Rodrigues, L. Rocha Barros Goncalves, L.R. Rodrigues, J.A. Texeira. Enzymatic synthesis of sugar esters and their potential as surfaceactive stabilizers of coconut milk emulsions. *Food Hydrocolloids*. **2012**, 27, 331.
9. L. Schiefelbein, M. Keller, F. Weissmann, M. Luber, F. Bracher, W. Friess. Synthesis, characterization and assessment of suitability of trehalose fatty acid esters as alternatives for polysorbates in protein formulation. *European journal of pharmaceutics and biopharmaceutics*. **2012**, 76, 350.
10. N.R. Kahn, V.K. Rathod. Enzyme catalyzed synthesis of costemic esters and its intensification: a review. A review. *Process Biochemistry*. 2015, 50, 1806.
11. C. N. Mulligan. Environmental applications for biosurfactants. *Environ. Pollution*. **2004**, 133, 198.
12. S.W. Hsieh, M.R. Lee, C.W. Tsai, L.Y. Lai, T. Yeh, C.W. Hsieh, T.J. Yang, S.W. Chang, Enzymatic synthesis, purification and identification of bioactive trehalose ester derivatives for health applications. *Food Bioprod. Process*. **2015**, 95, 172.
13. M. Habulin, S. Sabeder, Z. Knez. Enzymatic synthesis of sugar fatty acid esters in organic solvent and in supercritical carbon dioxide and their antimicrobial activity. *The Journal of Supercritical Fluids*. **2008,** 45, 345.
14. M. Ferrer, J. Soliveri, F.J. Plou, N. Lopez-Cortes, D. Reyes-Duartes, M. Christensen, J.L. Copa-Patino, A. Ballesteros. Synthesis of sugar esters in solvent mixtures by lipases from Thermomyces lanuginosus and Candida antarctica B, and their antimicrobial properties, *Enzyme and Microbial Technology*. **2005,** 36, 398.
15. T. Watanabe, S. Katayama, M. Matsubara, Y. Honda, M. Kuwahara. Antibacterial carbohydrate monoesters suppressing cell growth of Streptococcus mutans in the presence of sucrose. *Current microbiology*. **2000,** 41, 213.
16. J.K. Peterson, H.F. Harrison, O.T. Chortyk. Effects of various synthetic sucrose esters on weed seed germination and crop growth: structure - activity and dose - response relationships. *Journal of Agricultural and Food Chemistry*. **1997**, 45, 4837.
17. S. Okabe, M. Suganuma, Y. Tada. Dissacharide esters screened for inhibition of tumor necrosis factor - alpha release are new anti-cancer agents. *Japanese journal of cancer research*. **1999**, 90, 676.
18. C.C. Akoh, B.G. Swanson. Synthesis and properties of alkyl glycoside and stachyose fatty acid polyesters. *Journal of the American Oil Chemists ' Society*. **1989**, 66, 1301.

19. G. Paggiola. Biocatalysis in bio-derived solvents: an improved approach for medium optimisation, *Green Chemistry*. **2014,** 16, 2110.

20. P. Fickers, J. Destain, P. Thonart. Lipases are atypical hydrolases: main characteristics and applications. *Biotechnology agronomy, society and environment*. **2008,** 12, 130.

21. A. Dounia, G. Etienne, N. Stéphane, 6-O-glucose palmitate synthesis with lipase: Investigation of some key parameters. *Molecular Catalysis*. **2018,** 460, 68.

22. A. K. Chandel, R. Rudravaram, L. V. Rao, P. Ravindr, M. L. Narasu. Industrial enzymes in bioindustrial sector development: an Indian perspective. *Journal. Commercial Biotechnology*. **2007,** *13*, 283.

23. S. Nakamura. Using sucrose esters as food emulsifiers. *Inform* **1997,** 8, 874.

24. S. Cassel, C. Debaig, T. Benvegnu. P. Chaimbault, M. Lafosse, D. Plusquellec, P. Rollin. Original synthesis of linear, ranched and cyclic oligoglycerol standards. *European Journal of Organic Chemistry*. **2001,** 875, 1829.

25. N. Bouzaouit, C. Bidjou-Haiour. Response surface methodological study of glucose laurate synthesis catalyzed by immobilized lipase from candida cylindracea. *Biological Forum-An Intern. J.* **2016,** 81, 420.

26. S. Okazaki, M. Goto, H. Wariishi, H. Tanaka, S. Furusaki. Characterization and catalytic property of surfactant-laccase complex in organic media. *Biotechnology Progress*, **2000,** 16, 583.

27. V. Dandavate, D. Madamwar. Novel approach for the synthesis of ethyl iso valerate using surfactant coated *Candida rugosa* lipase immobilized in microemulsion based organogels. *Enzyme and microbial technology*. **2007,** 41, 265.

28. B. Eric, B. Jean-François. Enzymatic catalysis. *Techniques de l'ingénieur*, **1992,** 240, 1.

29. C. Didier, M. Pierre. Biocatalysis or enzymatic catalysis. *Techniques de l'ingénieur*, **2009,** 590, 1.

30. C. Blecker, M. Christophe, Immobilised lipases and their applications. *Biotechnology, Agronomy, Society and Environment*, **2008,** 12, 57

31. MF. Chaplin, C. Bucke. Enzyme Technology. *Cambridge University Press*, **1990.**

32. I. Alexander, S. Manfred. Methods for the immobilization of lipases and their use for ester synthesis. *Journal of Molecular Catalysis*, **1997,** 3, 303.

33. D. Sumitra, R. Christena, R. Yamuna, R. Sriramulu. Enzyme immobilization: an overview on techniques and support materials. *Biotech*, **2013,** 3, 1.

34. M. Eisenmenger, J. Reyes-De-Corcuera. High hydrostatic pressure increased stability and activity of immobilized lipase in hexane. *Enzyme and Microbial Technology*, **2009,** 45, 118.

35. Z. Dijkstra, R. Merchant, J.T.F. Keurentjes. Stability and activity of enzyme aggregates of *CALB* in supercritical CO_2. *Journal of Supercritical Fluids*, **2007,** 41, 102.

36. D. Combes, P. Monsan. Biocatalysis or enzymatic catalysis. *Techniques de l'ingénieur*, **2009,** 11, 1.

37. M.F. Chaplin, C. Bucke. Enzyme Technology. *Cambridge University Press*, **1990.**

38. J. M. Guisan. Immobilization of Enzymes and Cells. *Methods Biotechnol*, **2006,** 22.

39. U. Hanefeld, L. Gardossi and E. Magner, Understanding enzyme immobilisation. *Chemical Society Reviews* **2009,** 38, 468.

40. M.F. Chaplin, C. Bucke. Enzyme Technology. *Cambridge University Press*, **1990.**

41. M. M. Zheng, Y. Lu, L. Dong, P. M. Guo, Q. C. Deng, W. L. Li, Y. Q. Immobilization of Candida rugosa lipase on hydrophobic/strong cation-exchange functional silica particles for biocatalytic synthesis of phytosterol esters. *Bioresource technology*, **2012,** 115, 141.

42. A. Bastida, P. Sabuquillo, P. Armisen, R. Fernandez-Lafuente, J. Huguet and J. M. Guisan, A single step purification, immobilization, and hyperactivation of lipases via interfacial adsorption on strongly hydrophobic supports. *Biotechnology and bioengineering*, **1998,** 58, 486.

43. V, Iyer, S. Padma, L. Ananthanarayan. Enzyme stability and stabilization-aqueous and non-aqueous environment. *Process biochemistry*. **2008,** 43, 1019.

44. Klibanov, M. Alexander. Improving enzymes by using them in organic solvents. *Nature*, **2001**, 409, 241.

45. E. Brown, B. Jean-François. Enzymatic catalysis. Techniques de l'ingénieur, **1992**, 240, 1.

46. L. Ren, J. He, S. Zhang, D.G. Evans, X. Duan, Immobilization of penicillin G acylase in layered double hydroxides pillared by glutamate ions. *Journal of Molecular Catalysis B: Enzymatic*,, **2002**, 183, 11,

47. D. Combes, P. Monsan. Biocatalysis or enzymatic catalysis. *Techniques de l'ingénieur.* **2009**,11, 590.

48. K. R. Kaneto, A new tyrosinase biosensor based on covalent immobilization of enzyme on N-(3-aminopropyl) pyrrole polymer film. *Current Applied Physics*, **2005**, 5, 178.

49. E. Brown, B. Jean-François. Enzymatic catalysis. *Techniques de l'ingénieur*, **1992**, 240, 1.

50. B. Christophe. Immobilised lipases and their applications, *Biotechnology, Agronomy, Society and Environment*, **2008**, 12, 68.

51. D. Sumitra, R. Christena, R. S. R. Yamuna. Enzyme immobilization: an overview on techniques and support materials. *Biotech,* **2013**, 3, 9.

52. G. Iqbal, A. Ballesteros. Bioencapsulation within synthetic polymers (Part1): sol-gel encapsulated biologicals. *Tibtech,* **2000**, 18, 296.

53. M. Maciej, A. Krywko-Cendrowska, P. Krysinski, J. Rogalski. Encapsulation of laccase in a conducting polymer matrix: A simple route. *Synthetic Metals*, **2009**, 159, 1738.

54. L. Zhang, Y. Jiang, Sh. Jiafu, S. Xiaohui, Li Jian, Biomimetic polymer-inorganic hybrid microcapsules for yeast alcohol dehydrogenase encapsulation. *Reactive & Functional Polymers*, **2008**, 68, 151.

55. J.P. Chen, W.S. Lin, Sol-gel powders and supported sol-gel polymers for immobilization of lipase in ester synthesis. *Enzyme and Microbial Technology*, **2003**, 32, 801.

56. S. Fennouh, S. Guyon, C. Jourdat, J. Livage, C. Roux, Encapsulation of bacteria in silica gels. *Comptes Rendus de l'Académie des Sciences-Series IIC-Chemistry*, **1999**, 625.

57. G. Orive, R. M Hernandez, A.R. Gascon, M. Igartua, J. P. Pedraz, Microcapsules prepared with different biomaterials to immobilize GDNF secreting 3T3 fibroblasts. *International journal ofpharmaceutics*, **2003**, 259, 68.

58. R. Kumar, A. N Maitra, P. K. Patanjali *et al.* Hollow gold nanoparticles encapsulating horseradish peroxidase. *Biomaterials*. **2005**, 26, 6743.

59. P.V. Iyer, L. Ananthanarayan, Enzyme stability and stabilization: Aqueous and nonaqueous environment, *Process Biochemistry*, **2008,** 43, 1019.

60. N. Kamiya, M. Goto, F.Nakashio. Surfactant-Coated Lipase Suitable for the Enzymatic Resolution of Menthol as a Biocatalyst in Organic Media. *Biotechnologie. Progress*. **1995**, 11, 270.

61. K.P. Dhake, D.D. Thakare. B.M. Bhanage. Lipase: A potential biocatalyst for the synthesis of valuable flavour and fragrance ester compounds. *Flavour Fragr. J.* **2013**, 28, 71.

62. M.J. Ferrer, J.P. Francisco, N. López-Cortés, D. Reyes-Duarte, M. Christensen, J.L. Copa-Patiño, A. Ballesteros. Synthesis of sugar esters in solvent mixtures by lipases from *Thermomyces lanuginosus* and *Candida antarctica* B and their antimicrobial properties. *Enzyme and microbial technology.* **2005**, 36, 391.

63. A.M. Klibanov. Improving enzymes by using them in organic solvents. *Nature*, **2001**, 108, 241.

64. A. Zaks, A.M. Klibanov, Enzyme-catalyzed processes in organic solvents, *Proceedings of the National*

Academy of Sciences, **1985**, 82, 3192.

65. D. Feng-Xia, L. Zhang, T. Xiao-Zuo, Ch. Hong-Bing, W. Xiu-Li and W. Yu-Zhong, Ionic liquid coated lipase: Green synthesis of high molecular weight poly (1,4-dioxan-2-one). *Journal of Molecular Catalysis B: Enzymatic.* **2012,** 77, 46.

66. T. Mori, S. Kishimoto, K. Ijiro, A. Kobayashi and Y. Okahata. A Lipid-Coated Lipase as an Efficient Hydrolytic Catalyst in the Two-Phase Aqueous-Organic System, *Biotechnology and bioengineering.* **2001,** 76, 157.

67. N. Bouzaouit, C. Bidjou-Haiour. Response Surface Methodological Study of Glucose Laurate Synthesis Catalyzed by Immobilized Lipase from *Candida cylindraceaBiological Forum-An International Journal.* **2016,** 81, 420.

68. V. Dandavate and D. Madamwar, Novel approach for the synthesis of ethyl isovalerate using surfactant coated *Candida rugosa* lipase immobilized in microemulsion based organogels, *Enzyme and microbial technology.* **2007**, 41, 265.

69. Y. Okahata and T. Mori, Lipid-coated enzymes as efficient catalysts in organic media, *Trinds in Biotechnolgy.* **1997**, 4, 15.

70. N. Klai, C. Bidjou-Haiour. S. Bouquillon. D-Xylose-based surfactants: Synthesis, characterization and molecular modeling studies, *Comptes Rendus Chimie.* **2015**, 18, 599.

71. A.M. Klibanov, Improving enzymes by using them in organic solvents, *Nature*, **2001,** 108, 241-246.

72. N. Bouzaouit, C. Bidjou-Haiour. Response surface methodological study of glucose laurate synthesis catalyzed by immobilized lipase from candida cylindracea, *Forum-An International Journal.* **2016**, 81, 420.

73. M. M. Zheng, Y. Lu, L. Dong, P. M. Guo, Q. C. Deng, W. L. Li, Y. Q. Feng and F. H. Huang, Immobilization of Candida rugosa lipase on hydrophobic/strong cation-exchange functional silica particles for biocatalytic synthesis of phytosterol esters. *Bioresource technology*, **2012**, 115, 141.

74. T. Gitlesen, M. Bauer and P. Adlercreutz, Biochim. Adsorption of lipase on polypropylene powder. *Biochimica et Biophysica Acta (BBA)-Lipids and Lipid Metabolism,* **1997**, 1345, 188.

75. R. Fernandez-Lafuente, P. Armisen, P. Sabuquillo, G. Fernandez-Lorente and J. M. Guisan, Immobilization of lipases by selective adsorption on hydrophobic supports. *Chemistry and physics of lipids* ,**1998**, 93, 185.

76. A. Bastida, P. Sabuquillo, P. Armisen, R. Fernandez- Lafuente, J. Huguet and J. M. Guisan, A single step purification, immobilization, and hyperactivation of lipases via interfacial adsorption on strongly hydrophobic supports. *Biotechnology and* **bioengineering1998**, 58, 486.

77. M. Kreiner, B. D. Moore and M. C. Parker, Chem. Commun. Enzyme-coated micro-crystals: a 1-step method for high activity biocatalyst preparation, *Chemical communications.* **2001**, 1096-1097.

78. M. Kreiner, J. F. A. Fernandes, N. O'Farrell, P. J. Halling and M.-C. Parker, Stability of protein-coated microcrystals in organic solvents. *Journal of Molecular Catalysis B: Enzymatic.* **2005**, 33, 65-72.

79. J. J. Zheng, L. Xu, Y. Liu, X. Y. Zhang and Y. J. Yan, Lipase-coated K2SO4 micro-crystals: preparation, characterization, and application in biodiesel production using various oil feedstocks. *Bioresource Technology.* **2012**, 110, 224.

80. S. Shah, A. Sharma, M. N. Gupta, Cross-linked protein-coated microcrystals as biocatalysts in non-aqueous solvents. *Biocatalysis and Biotransformation*. **2008**, 26, 266.

81. M.A. Bradford, A rapid and sensitive method for the quantitation of microgram quantities of protein utilizing the principle of protein-dye binding, *Analytical biochemistry*. **1976**, 72, 248.

82. S.M. Gotham, P.J. Fryer, W. Paterson, The measurement of insoluble proteins using a modified Bradford assay, *Analytical biochemistry.* **1988**, 173, 353.

83. M. M. Bradford, A rapid and sensitive method for the quantitation of microgram quantities of protein utilizing the principle of protein-dye binding. *Analytical biochemistry*, **1976**, *72*, 248.

84. R. Reshmi, S. Sugunan, Superior activities of lipase immobilized on pure and hydrophobic clay supports:Characterization and catalytic activity studies, *Journal of Molecular Catalysis B: Enzymatic* **2013,** 97, 36.

85. L. F. Garcia, V. Gotor, Lipase-catalyzed transesterification in organic media: Solvent effects on equilibrium and individual rate constants. *Biotechnology and bioengineering*.**1998**, 59, 684.

86. M. Persson, I. Mladenoska, E. Wehtje, P. Adlercreutz, Preparation of lipases for use in organic solvents. *Enzyme and Microbial Technology*, **2002**, *31*, 833.

87. V.V. Mozhaev, I.V. Berezin, K. Martinek. Structurestability relationship in proteins: fundamental tasks and strategy for the development of stabilized enzyme catalysts for biotechnology, *Critical Reviews in Biochemistry*. **1988**, 3, 235.

88. Okahata Y. and Mori T., Lipid-coated enzymes as efficient catalysts in organic media, *Trends Biotechnology*, **1997**, 4, 15.

89. Y. Okahata, K. Ijiro. A lipid-coated lipase as a new catalyst for triglyceride synthesis in organic solvents, *Journal of the Chemical Society, Chemical Communications*. **1988,** 1, 1392.

90. K. Ramani, S. Karthikeyan, R. Boopathy, L. John Kennedy, A.B. Mandala, G. Sekaran. Surface functionalized mesoporous activated carbon for the immobilization of acidic lipase and their application to hydrolysis of waste cooked oil: Isotherm and kinetic studies. *Process biochemistry*. **2012,** 47, 435.

91. N. Rezgui, C. Bidjou-Haiour. Physical and chemical study of coated *Candida cylindracea* with sorbitan esters. *Research Journal of Chemistry and Environment*. **2021**, 25, 10.

92. K. Ramani. S. Karthikeyan. R. Boopathy, L. John Kennedy. A.B. Mandala, G. Sekaran. Surface functionalized mesoporous activated carbon for the immobilization of acidic lipase and their application to hydrolysis of waste cooked oil: Isotherm and kinetic studies", *Process biochemistry*. **2012,** 47, 435.

93. S. Gupta., S. Yogesh, S. Javiya, M. Bhambi, C.S. Pundir, K. Singh and A. Bhattacharya. Comparative study of performances of lipase immobilized asymmetric polysulfone and polyether sulfone membranes in olive oil hydrolysis. *International journal of biological macromolecules*. **2008,** 42, 145.

94. R. BenSalah, H. Ghamghui, N. Miled, H. Mejdoub, Y. Gargouri, Production of butyl acetate ester by lipase from novel strain of Rhizopus oryzae. *Journal of bioscience and bioengineering*. **2007**, *103*, 368.

95. a. P. Degn, W. Zimmerma, Optimization of carbohydrate fatty acid ester synthesis in organic media by a lipase from Candida antarctica. *Biotechnology and* **bioengineering2001**, *74*, 483.

b. Y. Li, D. Rethwisch, Scale-up of pseudo solid-phase enzymatic synthesis of α-methyl glucoside acrylate. *Biotechnology and bioengineerring*,. **2002**,*79*, 15.

96. P. Adlercreutz. Immobilisation and application of lipases in organic media. *Chemical Society Reviews*, **2013**, *42*, 6406.

97. M. K.Walsh, R. A Bombyk, A. Wagh, A. Bingham, L. M. Berreau, Synthesis of lactose monolaurate as influenced by various lipases and solvents. *Journal of Molecular Catalysis B: Enzymatic*, **2009**, 60, 171.

98. T. Kobayashi, Lipase-catalyzed syntheses of sugar esters in non-aqueousmedia, *Biotechnology. Letters*. **2011,** 33, 1911.

99. Z. Yang, Z.-L. Huang, Enzymatic synthesis of sugar fatty acid esters in ionic liquids, *Catalysis Science & Technology*. **2012,** 2 1767.

100. L. Xiao-Sheng, W. Qing, H. Ze-Lin, C. Yu-Zheng, J. Peter. Z. Hallingc, Impacts of ionic liquids on enzymatic synthesis of glucose laurate and optimization with superior productivity by response surface methodology. *Process Biochemistry*, **2015,** 50, 1852.

101. A. Ghanem, Trends in lipase-catalyzed asymmetric access to enantiomerically pure/enriched compounds. *Tetrahedron*, **2007**, 63, 1721-54.

102. L. Xiao-Sheng, W. Qing, H. Ze-Lin, C. Yu-Zheng, J. Peter. Z. Hallingc, Impacts of ionic liquids on enzymatic synthesis of glucose laurate and optimization with superior productivity by response surface methodology. *Process Biochemistry*, **2015,** 50, 1852.

103. X.Sh. Lina, Q. Wenb, Z.L. Huanga, C. Yu-Zheng, J. Peter. Impacts of ionic liquids on enzymatic synthesis of glucose laurate andoptimization with superior productivity by response surfacemethodology. *Process Biochemistry.* **2015,** 50, 1852.

104. S. Soultani, J-M. Engasser, M. Ghoul. Effect of acyl donor chain length and sugar/acyl donor molar ratio on enzymatic synthesis of fatty acid fructose esters. *Journal of Molecular Catalysis B: Enzymatic.* **2001**, 11, 31.

105. S. Hari Krishna, N. G. Karanth, Lipases and lipase-catalyzed esterification reactions in nonaqueous media. *Catalysis Reviews.* **2002,** *44*, 499.

106. M. M. Soumanou, U. T. Bornscheuer. Improvement in lipase-catalyzed synthesis of fatty acid methyl esters from sunflower oil. *Enzyme and microbial technology*, **2003**, *33*, 97.

107. S. Torres, G. R. Castro. Non-aqueous biocatalysis in homogeneous solvent systems. *Food Technol Biotechnol*, **2004,** *42*, 271.

108. J. Yu, J. Zhang, A. Zhao, X. Ma, Study of glucose ester synthesis by immobilized lipase from Candida sp. *Catalysis Communications*, **2008**,*9*,1369.

109. N. Li, D. Ma, M. H. Zong. Enhancing the activity and regioselectivity of lipases for 3'-benzoylation of floxuridine and its analogs by using ionic liquid-containing systems. *Journal of biotechnology*, **2008**, *133*, 103.

110. A.J. Carmichael, K.R. Seddon, Polarity study of some 1-alkyl-3-methylimidazolium ambient-temperature ionic liquids with the solvatochromic dye, Nile Red. *Journal of Physical Organic Chemistry*, **2000**, 13, 591.

111. L. Cao, U.T. Bornscheuer, R.D Schmid. Lipase-catalyzed solid-phase synthesis of sugar esters. Influence of immobilization on productivity and stability of the enzyme. *Journal of Molecular Catalysis B: Enzymatic.* **1999**, 6, 85.

112. Y. Sugimura, K. Fukunaga, T. Matsuno, K. Nakao, M. Goto, F. Nakashio. A study on the surface hydrophobicity of lipases, *Biochemical Engineering Journal,* **2000**, 5, 123-128.

113. D. Coulon, Girardin M.,Rovel B.,Ghoul M., Comparison of direct esterification and transesterification of fructose by *Candida* antarticalipase, *Biotechnology Letters*, **1995,** 17, 2, 183-186.

MIX
Papier aus verantwortungsvollen Quellen
Paper from responsible sources
FSC® C105338

Printed by Books on Demand GmbH, Norderstedt / Germany